Lotus in the Lab: Spreadsheet Applications for Scientists and Engineers

GW00722366

Glenn I. Ouchi

BREGO Research and
Laboratory PC Users Group

ADDISON-WESLEY PUBLISHING COMPANY
The Advanced Book Program
Redwood City, California · Menlo Park, California
Reading, Massachusetts · New York · Amsterdam
Don Mills, Ontario · Sydney · Bonn · Madrid · Singapore
Tokyo · San Juan · Wokingham, United Kingdom

To Gail, Rochelle, Bennett, Jean, and my Mom.

Publisher: Allan M. Wylde
Production Administrator: *Karen L. Garrison*
Editorial Coordinator: *Alexandra McDowell*
Electronic Production Consultant: *Mona Zeftel*
Promotions Manager: *Celina Gonzales*

Library of Congress Cataloging-in-Publication Data

Ouchi, Glenn I., 1949–
  Lotus in the lab.

  Includes index.
  1. Science—Data processing. 2. Engineering—Data processing. 3. Lotus 1-2-3 (Computer program)
I. Title.
Q183.9.093 1988   005.36'9'0245   87–22960
ISBN 0–201–14349–6
ISBN 0–201–14307–0 (pbk.)

ABCDEFGHIJ-AL-898

Preface

This book describes the skills and techniques needed to use Lotus 1-2-3 as a data analysis and reporting tool in a scientific or engineering laboratory. Since its introduction, the focus for applications of Lotus 1-2-3 has been predominately for business. Fortunately, the data analysis and reporting capabilities of 1-2-3 have not been overlooked by scientists and engineers. Surveys have found from 10-20 percent of Lotus 1-2-3 users are scientists or engineers. No doubt, many of these engineers and scientists are using 1-2-3 for "classical business" applications, to create budgets and run their businesses or departments. But they are also using 1-2-3 to analyze and report their laboratory data.

Using the applications in this book you can put Lotus 1-2-3 to work solving your laboratory problems. Lotus 1-2-3 is also an excellent program for building mathematical models. In just a few minutes, you can develop a model and test its results. The results can be graphed for even easier data analysis. Then the model and data can be stored on disk for future analysis or archival storage.

How This Book Is Organized

This book is divided into two parts, basic techniques for utilizing 1-2-3 for scientists and engineers and laboratory applications. Part One, Basic Techniques, contains information about Lotus 1-2-3, how you can use it in your laboratory and how to import and export your experimental data in and out of the program. Chapter one is an introductory chapter describing what 1-2-3 is and why it is a valuable tool for scientists and engineers. Chapter two, Capturing Experimental Data, describes the many ways laboratory data can be captured on disk for analysis with 1-2-3. This chapter includes step-by-step instructions on how to use RS-232 and IEEE-488 interfaces and a data acquisition card to acquire data from your laboratory instruments. Chapter three, Importing Data into 1-2-3, describes how data on disk can be loaded into 1-2-3. Chapter four, Presenting Your Results, describes how your data and analysis results can be presented and communicated to others. Programs which work directly with Lotus 1-2-3 data and graphs to produce convincing documents, reports, and presentation aids are described in this chapter.

These chapters are followed by Part Two, Laboratory Applications, where each chapter describes an application you can perform using 1-2-3. The first four are short introductory applications which can be performed in just a few minutes. These applications show the basic capabilities of 1-2-3. These applications include: Testing the Precision and Dynamic Range of the numbers you can use in 1-2-3, creating a Pass/Fail Report from analysis results, reporting the mean, standard deviation, high, and low values from experimental data and sorting and graphing data for data analysis.

Then there are ten advanced laboratory applications. These applications include: Plotting and maintaining control chart data, generating and plotting a frequency distribution from experimental data, loading and summarizing data from many different files, using data regression to fit linear, quadratic, and cubic functions to data and plotting the results, plotting confidence intervals for a calibration curve, fitting and plotting an exponential function to data, creating and using a data analysis log book, creating a multi-component report, and plotting semi-log plots.

1-2-3 novices should start by reading the first chapter of this book. Then get some hands-on experience by working through the Lotus 1-2-3 tutorial. Once you know the basic skills try performing the four introductory applications in the applications section. After completing the introductory applications, read chapters two and three. Then work your way through the advanced applications.

For those already familiar with 1-2-3, you can start by reading the first three chapters. Then you can look over the list of applications and select those which are the most interesting to you. After working through those applications, you can work through the balance of applications to learn other new skills and techniques.

Other Spreadsheets

A number of new spreadsheet programs have been recently introduced. Two major introductions, Excel by Microsoft and Quattro by Borland International should add even more momentum to the rapidly expanding use of spreadsheets. The techniques and applications described in this book can be performed in these new products since they have 1-2-3 compatibility.

Lotus 1-2-3 Student Version

Lotus Development and Addison-Wesley have developed a student version of Lotus 1-2-3. This version of 1-2-3 has all of the functionality of the regular version of 1-2-3 except it has a smaller worksheet and saves different worksheet files. The student version has only 64 columns and 256 rows compared to 256 columns and 8192 rows in the regular version of 1-2-3. All of the applications in this book can be performed with the student version of 1-2-3. In fact, since the student version of 1-2-3 can retrieve regular 1-2-3 files, the disk which is offered with the completed applications (see below) can be used with the student version.

This book along with the student version of 1-2-3 can be used in a course for teaching computerized data acquisition and data analysis to science and engineering students. Lectures can be developed to cover the material in Part One, Basic Techniques, and laboratories can be based on the applications presented in Part Two, Laboratory Applications.

Applications Available on Disk

All of the completed applications in this book are available on a disk, ready to be loaded and utilized with your copy of Lotus 1-2-3. To order a copy of this disk, fill out the disk order card at the back of this book and send it to the address with the fee. Each application on disk includes two files, one containing the original experimental data, and the second with the completed application. Starting with the experimental data, you can work through an application as it is described in the book. That is the best way to learn the skills to use Lotus 1-2-3.

Acknowledgements

Finally, I would like to acknowledge the people who helped make the publication of this book possible. They are: Allan Wylde, Karen Garrison, Mary Ellen Holt, Jacqueline Davies, Joseph Priest, Jim Tung, and Celina Gonzales. My special thanks goes to Allan Wylde for his help in getting this book published.

Glenn I. Ouchi, Ph.D.
San Jose, CA
March 1988

Typographical Conventions

1-2-3 commands and instructions are issued by pressing specific sequences of keys. Keystroke sequences in this book are designated by symbols for arrow keys, and function keys. Named keys like [Esc], [Home] and [End] are enclosed in brackets. The following terminology will be used in the book:

Press: This means to strike or press a key. It is used primarily with the arrow keys, special keys, function keys and the enter or return key [Enter], to perform an operation. For example, the instruction to press the [Enter] key would appear as shown below.

>   **Press:** **[Enter]**

Type: Type also means to strike or press a key. It is used primarily with the standard typewriter keys to enter information into the worksheet. Keys to be typed are boldfaced. For example, the instructions to type the word Calibration would appear as shown below.

>   **Type:** **Calibration**

Move to: This means to move the cell pointer or cursor to a specific location on the worksheet. This is usually done with a combination of arrow keys or other function keys. For example, the instruction to move the cell pointer to cell G5 would appear as shown below.

>   **Move to: G5**

Combination Keystrokes

If two keys are used in combination to perform a single task, they are presented on the same line. Hold down the first key while the second key is pressed. For example, the instruction to hold down the ALT key while pressing the letter k would appear as shown below.

>   **Press: [ALT] k**

Command Sequences

Command sequences in this book are offset and indented. The first letter of each command option is boldfaced, indicating that it should be typed or the command can be selected from the menu. Any other parts of the command sequence to be typed are also boldfaced. For example, the command to retrieve the worksheet file named ACTIVE is shown below.

```
/File Retrieve ACTIVE [Enter]
```

In this example you would type the characters **/FRACTIVE** and then press [Enter].

Cell Contents

The contents of cells are displayed in the same form used by 1-2-3 when displaying the contents of a cell in the control panel. For example,

```
B4: +A4/300
```

B4 is the cell identifier which contains the entry +A4/300. To enter the contents of this cell

```
Move to: B4
Type:  +A4/300
Press: [Enter]
```

Function, File and Range Names

Function names will appear as all capitals as in @SUM(A1..A5) and @AVG(A1..A5). File and range names when entered as part of a command sequence will also appear as all capitals as in:

```
/Range Name Create THIS_DATA [Enter]
```

However, when using a file or range name as part of a macro, lower case letters will be used to distinguish these names from single letter commands as in the following macro step:

```
/RNCthis_data~
```

Contents

PART ONE: BASIC Techniques

1

Introduction to Lotus 1-2-3

Lotus 1-2-3 integrates in a single program three major types of application software: spreadsheets, graphics and data management. 1-2-3 is operated by selecting commands from easy to use menus or by chaining together menu commands using the Lotus Command Language in Lotus macros. Although 1-2-3 was initially targeted for business users, it can just as easily be used to perform applications for scientific and engineering problem solving. 1-2-3 provides the user with so many computing capabilities, it has proven to be as valuable a tool in the laboratory as it has been in the finance department.

This book will introduce you to 1-2-3 by providing a series of applications you can put to work right now in your own laboratory. The regular 1-2-3 documentation will teach you the mechanics of executing commands and their function. This book will provide you with complete applications and techniques you can perform by putting together a series of 1-2-3 commands to solve scientific and engineering problems. Whether you are a 1-2-3 novice or expert you will find these applications interesting and by performing them yourself, you will broaden your knowledge about using 1-2-3 and computers in your research.

You will find many of the applications are similar to the ones you may have seen solved in a programming language book for scientists and engineers. Just as problems can be solved by writing a program with a programming language like BASIC, C or Pascal, you can solve your laboratory problems by creating a Lotus 1-2-3 worksheet. You will be able to solve specific problems, which once could only be solved by writing a program, simply by entering data and executing commands in 1-2-3.

Each application will be introduced by describing the problem 1-2-3 will solve. These problems are general in nature and the solutions will be easily extended to be useful in many disciplines. You will no doubt find similar problems in your work. Think of this book as a series of problem solving methods all based on using Lotus 1-2-3.

Extending Your Computing Capabilities

By learning the techniques in this book you will expand your knowledge on how to use powerful problem solving tools. A program like 1-2-3 with your personal computer provides a lowcost "software toolbox" for solving problems.

These tools give us the ability to model or simulate physical phenomena. By building a computer model, we can test the model hundreds or thousands of times in just a few minutes. Time consuming and expensive physical models in many cases need not be created because computer models can be created and tested.

These tools remove the burden of having to write programs to peform calculations. This allows us to focus on "strategies" for finding the best solution rather than the "mechanics" of getting a solution. The tools free us to experiment with many potential solutions in a short period of time and we can select the best solution. Only with the aid of a program like 1-2-3 can this "what-if" analysis be easily performed.

1-2-3 simulations can also be used to teach. In most cases, in a modern laboratory we interact with the "real-world" through instruments to collect our data. Computer controlled instruments acquire and store data from various experiments. 1-2-3 can be used to teach data analysis and data interpretation skills. Basic analysis skills and techniques can be taught in this way without expensive instrumentation.

Why is Lotus 1-2-3 So Popular in the Lab?

Lotus 1-2-3 is popular in the laboratory because it provides all the basic computing tools to create, analyze and test mathematical models using its spreadsheet, graphics and data management capabilities. By writing Lotus macros and utilizing the Lotus Command Language, manual operations can be automated to provide a complete turn-key application. 1-2-3 can be further extended by using one or more add-in or add-on programs. These programs provide additional commands and capabilities not found in 1-2-3 such as word processing, 3D graphics, forms for data entry, data storage compression and direct data acquisition.

Let's look at each of the basic parts of 1-2-3 to see why with your computer it makes such a powerful computing tool for your laboratory.

1-2-3 Spreadsheet

The 1-2-3 spreadsheet can perform any task you have previously performed with paper, a pencil and a calculator. Using 1-2-3, the computer screen becomes your paper, the computer keyboard is your pencil and the computer performs the calculations. Even more convenient is that you can store the contents of an entire worksheet to work on later or archive the information. Stored data can also be consolidated into a single summary report. Once you have your data in computer readable form, you have many ways to analyze and report your data. The data can be easily passed to other programs or other computers.

There are three major elements of the 1-2-3 spreadsheet which makes it popular and useful as a laboratory data handling tool. First, the screen setup of rows and columns resembles the grid you find on the pages in most laboratory notebooks. Second, the "What-You-See-Is-What-You-Get" or WYSIWYG philosophy allows direct construction of exactly the report you want to print. There are no abstract requirements to enter data or labels, you simply move to the location you want and type in the entry. What you see on the screen is the report you will get on your printer or disk file. Finally, the program gives instantaneous feedback with every entry. Worksheet values are recalculated and all values which are dependent on a newly added value are updated. If you do not like how a column of values in a report are formatted you can reformat the column. If you make an erroneous entry, you see it immediately and you can correct it.

The electronic worksheet is organized as a grid of columns and rows (Figure 1.1). Each column is identified by a letter (A,B,C,D...) and each row is labelled with a number (1,2,3,4...). At each intersection of a row and column are entry positions or cells which can contain information. Each cell is identified by the column letter and row number which intersect at the cell's location.

There are five types of information which can be placed in a cell:

1. A Number (e.g. 123, -12.34 or .09876)
2. Text or a Title (e.g. ACME Research & Development Labs)
3. A Formula (e.g. 100*C3, multiply the contents of cell C3 by 100)
4. A pre-programmed function (e.g. @AVG(range or cells) display the average of the given range of cells)
5. Commands for a macro.

```
        A         B         C         D         E         F         G
 1                Liquid Chromatograpy Data
 2                Area Percent Report
 3      =====================================================
 4      Sample:   QA #2345           Date:       04-Mar-85
 5      Column:   Sephadex           Operator:GIO
 6      Solvent:  Methanol/Water     Inst No.:LC #1234
 7      Notes:    Sample vial seal broken during shipment
 8      =====================================================
 9      Peak No. Ret. Time   Area    % Total Area
10             1     2.34    12345      1.160
11             2     2.67    34567      3.248
12             3     3.68    78904      7.414
13             4     4.98     5674      0.533
14             5     6.89    12356      1.161
15             6     9.78   896654     84.249
16             7    10.76     9223      0.867
17             8    13.45    14568      1.369
18                          ------------------
19                          1064291   100.000
20
```

Figure 1.1 Typical Lotus 1-2-3 worksheet with data entered into columns and rows, just like in a laboratory notebook.

To enter information into the worksheet you simply move a lighted box (called the cell pointer or cursor) to the grid location you wish to use and type in the entry. With numerical data you can perform the basic math functions (addition, subtraction, multiplication and division) and many other functions are preprogrammed for easy execution. Included in the pre-programmed functions are basic statistical calculations like average, standard deviation and variance. One of the real powers of 1-2-3 is when you can change a value in the worksheet, all values which are computed from that particular value are instantly recomputed and the results displayed on the worksheet. Using this capability, you can perform "what if" calculations on your models in just a few moments.

The spreadsheet also features the ability to copy or move the contents of one cell to other cells. This feature allows you to create an equation in one cell and then without having to re-enter the contents of the cell, use the same equation at another location in the worksheet. Moving cells allows you to display your data exactly the way you wish. If you have already entered a column of values and then find they are located in the wrong place on the worksheet, you can easily move them.

Using these simple rules complex models can be developed and utilized. Since you can store the entire worksheet you can build "templates" of your calculations and reports. Once you have prepared a report template you or anyone else can simply key in new raw data values and a new report from the current data will be completed.

Lotus Graphics

Graphs have always been an important tool for laboratory data analysis and data presentation. Generating graphs by hand is a tedious task. With 1-2-3, graphs can be generated from any data you have on your worksheet. Once graph parameters have been set, a graph of the current data can be generated with the touch of one key, the [GRAPH] key (which is usually [F10], function key 10). The current values of all parameters will always be graphed so you can change values on your worksheet and then instantly see the results in graphic form on your screen. This is another form of "what-if" analysis you can perform, except this time you can review graphs rather than columns of numbers. You can interactively test a model and see all of your results in graphical form. You will find this to be a valuable problem solving tool.

	A	B	C	D	E	F	G	H
12	From This:							
13	NAME	MG/ML		RT	AREA BC			
14								
15	COMP1	304.		4.33	132364 01			
16	COMP2	202.		8.54	542382 01			
17	COMP3	683.		11.46	789383 01			
18	COMP4	105.		17.29	678342 01			
19								
20	TOTALS	1294.			2142471			
21								
22	To This:							
23	NAME	MG/ML	RT	AREA	YLD FAC	YIELD	PASS/FAIL	
24	COMP1	304	4.33	132364	4	1216	PASS	
25	COMP2	202	8.54	542382	3.4	686.8	PASS	
26	COMP3	683	11.46	789383	5.6	3824.8	FAIL	
27	COMP4	105	17.29	678342	8.6	903	PASS	
28								
29								

Figure 1.2 Analysis reports can be reformatted and values recalculated.

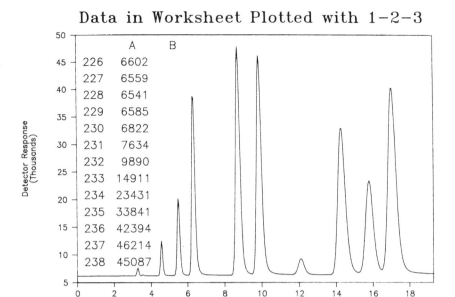

Data in Worksheet Plotted with 1-2-3

Graphs are an excellent way to truly visualize your data, easily see trends, and make convincing presentations of your data and data analysis. All graphs can be plotted directly on dot-matrix printers, plotters, laser printers and even slide making machines. The graphs can also be combined with text using other software products to make complete reports and documents.

Lotus Data Management

Data management commands can be performed on the worksheet data. A database can be set up in 1-2-3 and data within the database can be sorted, queried, extracted and found using Lotus data management commands. 1-2-3 also has commands to perform frequency distributions and statistical calculations on worksheet data.

1-2-3 will not make you think about retiring a good standalone data management program, but the commands available are very powerful and useful for many applications. They are particularily powerful when used in conjunction with 1-2-3 graphics and spreadsheet calculations. Many scientists use 1-2-3 in conjunction with a standalone data management programs like dBASE III Plus, R:Base or Paradox2. These programs are excellent database management programs but they do not have interactive calculation and graphing capabilities. Users can unload chunks of data stored in the data management program into a file which can be read into 1-2-3 for interactive data analysis and presen

	A	B	C	D	E	F
5	Data Log					
6	==					
7	Ref no.	Rec Date	Submitted by	Analysis	Analyte	Raw Value
8	101	8308.01	Smith	GCHROM	propane	201045
9	102	8308.01	Jones	LCHROM	asprin	45346
10	103	8308.02	Green	ICP	Hg	2314560
11	104	8308.03	Brown	ICP	Hg	2150780
12	105	8308.03	Johnson	UV	asprin	35245
13	106	8308.04	Edger	IR	asprin	40563
14	107	8308.04	Johnson	GCHROM	butane	23568
15	108	8308.04	Weaver	GPC	polyv	342789
16	109	8308.05	Mills	ICP	Hg	1894670
17	110	8309.06	Peterson	GPC	vinyl	3456790
18	2001	8307.05	Calibration	GCHROM	propane	230145
19	2002	8307.05	Calibration	GCHROM	methane	150032
20	2003	8307.05	Calibration	GCHROM	ethane	120132
21	2004	8307.12	Calibration	GCHROM	propane	234024
22	2005	8307.12	Calibration	GCHROM	methane	160054

Figure 1.4 Data can be stored in a database, sorted, and data meeting specific criteria extracted from the database or found and highlighted on the screen.

tation. Any data entered into 1-2-3 can be easily loaded into the standalone data management programs without having to retype the data. By using the programs in this way, you will get the most out of both program types.

For computer novices, the data management commands in Lotus 1-2-3 provide a good introduction to data management techniques and terminology. The data management skills you will learn using 1-2-3 can be carried over to other more powerful data management programs.

Lotus Macros

The mechanics of data entry, analysis and reporting can also be automated by using Lotus "macros". Macros allow you to assign several key strokes to a single keystroke. Macros also have programmable steps so logic can be built into the macro execution. Macros are powerful automation tools because they can automate a predominately manual application. With a macro, you can make all interaction with a worksheet totally automatic including making your own application specific menus. A Lotus novice can perform an application by pressing a single keystroke combination and then simply select menu choices and answer prompts rather than having to make all the right 1-2-3 menu selections and place data in the proper cell locations.

Lotus macros are built from a combination of keystrokes and Lotus Command Language (LCL) operations. The LCL provides all of the essential program control commands to create programs within 1-2-3. For example, the commands to perform a BASIC "FOR- NEXT" loop with a subroutine call would be:

```
FOR counter = 1 TO 10
    GOSUB get_data
NEXT
```

In this example **counter** is a variable and **get_data** is a subroutine name.

The same program structure could be performed in a Lotus 1-2-3 macro with the command:

```
{FOR counter,1,10,1,get_data}
```

Here **counter** is a range-name which designates a cell on the worksheet which counts the number of times the **get_data** subroutine is called. The subroutine get_data is also a range-name identifying the starting cell of commands in the subroutine.

The complete mechanics of how macros are created and executed will be covered in Part Two.

Lotus Add-ons and Add-ins

With literally millions of users of Lotus 1-2-3, software developers have created products which work with and from within the popular program. These additional programs perform functions not found in Lotus 1-2-3.

Programs that work with Lotus 1-2-3 are called add-on programs. Examples of these programs are Lotus Freelance Plus, SQZ! from Turner-Hall and Sideways from Funk Software. These programs work with 1-2-3 data files outside of the actual 1-2-3 program itself.

Lotus Freelance Plus can read Lotus PIC files and your graphs can be improved, edited and combined with other symbols or freehand drawings. Worksheet files can be read and data turned into graphs. Thus the simple graphs which you can create in 1-2-3 can be improved for more effective presentations.

SQZ! can compress the 1-2-3 Worksheet files so they take up less space on your disks. The SQZ! program is loaded into memory just before you load 1-2-3. When you want to retrieve or save a worksheet file, SQZ! will be activated. If you are saving a file, the file will be compressed so it will take up much less room on the disk. Compressions of up to 90 percent can be made. The resulting worksheet file is given the extension ".WK!" to indicate it is a SQZ! file rather than the normal extension. When you want to retrieve a SQZ! file, the data is unsquezzed to obtain the original data.

Sideways prints 1-2-3 ".PRN" files on printers from top to bottom rather than from right to left. This allows wide worksheets to be printed sideways down the paper. There are now two versions of Sideways, the original version is an add-on program which works only with 1-2-3 files stored on disk. The newest version of Sideways is an add-in program which can print data sideways directly from within Lotus 1-2-3.

Add-In Programs

Lotus add-in programs can directly manipulate the data in the worksheet. Even Lotus 1-2-3 experts can not perform every type of application they wish they could. Add-in programs can provide functions and features not found in the original 1-2-3 program, like direct data acquisition from instruments, three dimensional graphics and wordprocessing. These programs are created with programmer's tools sold by Lotus to program developers. A programmer can purchase the "Lotus Developer Tools" and then write add-in programs or add-in functions in assembly language. With the add-in program capability, Lotus 1-2-3 is acting much like an operating environment where programs developed by other companies can be executed while still using all the Lotus facilities.

Add-in programs can also provide capabilities not needed by all Lotus users but very beneficial for a specific group of 1-2-3 users. A large number of Lotus add-in programs are now available. Good examples of add-in programs providing capabilities of interest for laboratory use are Lotus Measure from Lotus Development, 4Word from Turner-Hall and 3D Graphics from Intex.

Lotus Measure

Lotus Measure was one of the first Lotus add-in products. Measure provides 1-2-3 with a direct interface to the real-world. Data can be acquired and sent in and out directly from a worksheet. Instruments with RS-232 or IEEE-488 ports can send and receive data directly with a 1-2-3 worksheet. Measure can also collect data acquired using a MetraByte DAS-16 card. This plug-in card has analog-to-digital converters, digital-to-analog converters and input/output lines.

Lotus Measure provides these capabilities by adding new commands to the 1-2-3 menus and Lotus Command Language. These commands can be utilized in one or more Lotus macros to perform data acquisition or control tasks. With Measure, fully automated applications including data acquisition, instrument control, and data analysis can be performed within 1-2-3.

4WORD

A basic application not provided by Lotus 1-2-3 is wordprocessing. This makes wordprocessing a perfect application for a Lotus add-in program. 4WORD is an add-in wordprocessor allowing you to create letters, memos and documents right inside 1-2-3. Some of the wordprocessing features provided by 4WORD include automatic word-wrap, block copy and move, bold, italics, and underline. 4WORD can also print form letters using a 1-2-3 database with its mail-merge capability. 1-2-3 charts and graphs can also be printed with the text you have entered into 4WORD. Turner-Hall also distributes Spellin! a spelling checker for Lotus 1-2-3. See additional information about add-in wordprocessors in the "Presenting Your Results" chapter.

3D Graphics

3D Graphics can plot data in three dimensions. This capability is another excellent example of a feature provided by an add-in program which is not provided by Lotus 1-2-3. With 3D-Graphics you can generate 3D line and bar charts. The resulting graphs are stored in regular "PIC" file format so they can be printed or plotted using the PGRAPH program. Up to a 100-by-100 matrix of data can be plotted with this add-in program. A complete description of how to use 3D Graphics is provided in the Applications section.

The Future of Lotus 1-2-3

Since its introduction in 1983, Lotus 1-2-3 has been the top selling program on the IBM PC. New versions of the integrated spreadsheet, graphics and data management program have been announced for the IBM PS/2 and other computers which run the OS/2 operating system. A new graphics based version, 1-2-3/G, has also been announced to take advantage of the OS/2 graphics based user interface. Versions of the program are also being developed to be run on IBM mainframes and no doubt many other popular computers.

Lotus has also announced a product called Lotus Extended Application Facility (LEAF). LEAF promises to allow Lotus applications to exchange real-time data among themselves. This will provide OS/2 users the type of applications software which can take advantage of the new hardware and operating system. LEAF should allow for example, a change in a spreadsheet application value to instantly update data in a database program and a graphics program. Watch for Lotus to also roll out a number of new applications programs in each major software market area all held together by LEAF.

New versions of 1-2-3 will no doubt include more commands and features to make the program an even more powerful tool. With the program available on so many different computers, it should become so common a program, nearly every computer user will be familiar with its use and applications. The techniques you will learn in this book will carry over to the next generation of 1-2-3 no matter what computer it is running on.

2

Acquiring Experimental Data

Data you use in 1-2-3 can be acquired from scientific instruments, other computers or directly from your experiments using a data acquisition interface card.

Where will the data you use in 1-2-3 come from? If you are like many 1-2-3 users, the data will be entered through your keyboard. But manual data entry is both time consuming and prone to typing and omission errors. The amount of data you can utilize for analysis is severely limited if you rely on hand entry. You should attempt to capture the data you wish to analyze and load it into 1-2-3 in an easy, reliable manner.

Many laboratory instruments use the same computer you will be using to run 1-2-3 as an instrument controller, data acquisition and data handling device. Data is stored directly on disk for analysis and reporting. Using this data in 1-2-3 requires knowing how to load the data from disk into the 1-2-3 program. If your applications fall into this catagory, you can go directly to Chapter 3 "Importing Data into 1-2-3".

If your laboratory data is not yet on your PC's disk then read on. We will explore three ways you can get your experimental data on to disk. First we will investigate how to use two communication ports, RS-232 and IEEE-488. Then we will see how to use a data acquisition card to acquire data.

During this discussion we will make specific references to an instrument, a gas chromatograph, which is representative of most instruments you will find in a laboratory. We will use this instrument and its data to describe basic data acquisition techniques and strategies. Even though the instrument you wish to acquire data from is not a gas chromatograph, you will be able to extend what you learn to many other instruments.

Acquiring
Chromatography
Data

A chromatograph is an instrument which can perform a simple but powerful task. It can separate a sample mixture into its individual chemical compounds (components), Figure 2.1. Chromatographs are widely used in the chemical, petrochemical, pharmaceutical, environmental, medical and food industry to obtain both quantitative and qualitative data about chemicals and biomolecules in mixtures. The sample mixture is injected onto a column which is internally coated or packed with special materials called column packing. The components in the mixture are separated as they migrate through the column because they have different interactions with the stationary column packing. Some components have very little interaction with the column packing and pass rapidly through the column while other components have many more interactions and move more slowly. The components are propelled through the column by a gas whose flow is precisely regulated. Another type of chromatograph is a liquid chromatograph. In a liquid chromatograph the components are carried in a liquid.

Figure 2.1 Block diagram of a gas chromatograph. A gas chromatograph has four main components: an injector where the sample is vaporized, a column where separation occurs, a detector where components are detected and an oven to control the temperature of all the components.

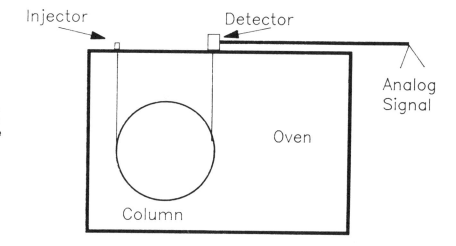

A Gas Chromatograph

In a gas chromatograph the sample must be vaporized so it can migrate through the column in gas phase. In a liquid chromatograph, the sample migrates down the column in liquid phase. As the components emerge from the end of the column, they are monitored by a detector.

The detector generates a voltage which rises and falls as a component emerges from the column. This produces a "peak" whose area is proportional to the original amount of the component in the sample mixture. Voltages are called analog signals since they are continuous and can be directly measured. Components can be identified by the amount of time required to migrate down the column under specific conditions. This time period, measured from the time the sample is injected until the component reaches the detector, is called the retention time. Retention times are found by running a pure standard of the compound of interest under the same conditions as the sample. Under the same conditions, the retention times of a component are very reproducible using modern chromatographs.

The simplest way to monitor the voltage changes generated by the detector is to use a strip-chart recorder. The strip-chart recorder pen will trace the voltage changes onto paper, making a record of the detector output. The peak areas and retention times would then have to be measured with a ruler by hand. With potentially hundreds or even thousands of peaks generated in a single experiment, manual data analysis from a strip-chart recorder is unpractical since it requires too much manual labor.

To replace a strip chart recorder, instruments called "digital integrators" were developed. A digital integrator (sometimes called just an integrator) acquires the voltage by digitizing the signal using an analog-to-digital converter. Data analysis programs running in the integrator then compute the areas under the peaks and compute their retention times. Some chromatographs include integrators as a built-in module. There are also chromatography data systems which use a PC fitted with the proper data acquisition hardware and software which perform the same functions as an integrator.

Raw Data and Results

Like many experiments, there are two levels of chromatography data. The first level is the actual digitized voltage signal which is called raw data. If this data is acquired properly, a plot of this data will look nearly identical to a strip-chart recording of an analysis. With this data, the experimental results can be reconstructed at any time. There is a large volume of this data. A typical 30 minute experiment

Figure 2.2 Data ac-
quired directly from a
transducer or sensor us-
ing a data acquisition
card.

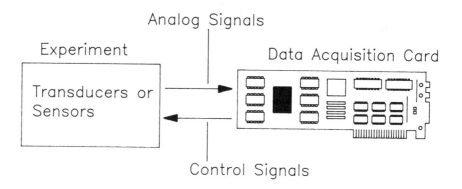

sampling the detector volotage at a data point every second would produce 1800 data points.

The second level of data are the areas and retention times found by
analyzing the digitized voltage signal. Here there is less data since
only the area and retention time for each peak found in a run needs to be
stored.

Chromatographs and the instruments which acquire and store their
data will provide many concrete examples for techniques, and strategies described in the rest of this chapter and in the applications section
of this book.

Evaluating Your Situation

First you must identify the source of your data. Normally this falls into
one of two broad catagories:

1. Direct readings from the experimental apparatus, using transducers or sensors as shown in Figure 2.2. Many laboratory instruments fall in this category.

   Transducers and sensors convert physical parameters like pressure, temperature, position, velocity, and light into an electrical
   signal for the purpose of measuremnt and control. For example, to
   measure the temperature changes during an experiment a thermocouple is used. A thermocouple will convert the temperature
   into a voltage signal. The voltage is an analog (continuous) signal which must be digitized to be used by a computer.

   In our chromatography example, the detector is the transducer
   since it converts the presence of a chemical compound passing
   through the detector into a voltage. If your data are generated in
   this category, you will need to use an interface card or another

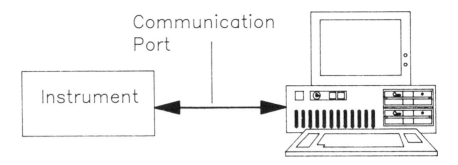

Figure 2.3 Data acquired from an instrument. The instrument acquires the data, converts it to digital form and then sends the digital data out the communications port. This instrument has an RS-232 communications port. The PC can acquire the data through an installed RS-232 communications port.

type of data acquisition unit to acquire the data and store the data on disk. An example of how an interface card is used is described in the "Using a Data Acquisition Interface Card" section found later in this chapter.

2. Instruments which acquire the data and can transmit the data out a communications port as shown in Figure 2.3.

The most common communication ports found on laboratory instrumentation are termed RS-232 and IEEE-488. A RS-232 port is also called a serial port or an asynchronous communications port. IEEE-488 is also called GPIB (General Purpose Interface Bus) and HPIB (Hewlett-Packard Interface Bus).

Chromatography integrators and built-in data acquisition modules usually have one of these communication ports. If your data source falls into this category, you can acquire data from the instrument through the communications port. If the instrument has an RS-232 or IEEE-488 port, you can follow the steps for that port described in this chapter. If the instrument does not have a data communications port, ask the instrument vendor if it is possible to add a communications port. Many instruments have these ports as an option. If no communications port is available, you may be able to acquire the data directly using a data acquisition interface card as described above in category 1.

General data acquisition techniques and strategies will now be discussed.

Data Acquisition Strategies

Data acquisition from instruments and data analysis can be placed into one of three categories Figure 2.4. The first category is real-time operation. Real-time means as the event occurs. In real-time operation, data is analyzed just as it is acquired. This type of operation may be needed so a decision or a control mechanism can be actuated immediately or automatically. In many cases this type of operation was implemented simply because the instrument manufacturer at the time the instrument was developed could not afford to include the necessary memory or mass storage device and thus there is no place to store the original data.

Real-time data acquisition and analysis is the most demanding category because data acquisition and analysis are closely linked. In our chromatography example, most chromatography integrators perform real-time detection of peaks and computation of areas and retention times. As the data is collected, a peak detection and integration algorithm is performed on the incoming data. Using this strategy, only a few hundred data points need to be in memory at any time. By performing real-time data analysis, only retention times and areas need be stored.

Real-time operation is fine if the data analysis algorithms are simple and have little chance of making an error. However, if more complex data analysis is needed, where for example human judgement should be used, a less demanding form of data acquisition and analysis should implemented. In chromatography for example, many data systems now store the raw data. This allows users to perform data analysis at a future time, not just as the data is collected. This can be very beneficial if there are any questions about how the data analysis was performed.

	Acquisition	Analysis
Real-Time	Automatic	Automatic
Semi-Automatic	Automatic	Command
Command	Command	Command

Figure 2.4 Three application categories for data acquisition and data analysis. The less automatic the data acquisition and analysis, the easier the application will be to implement. Automatic data acquisition directly into Lotus 1-2-3 requires the use of the add-in program Lotus Measure.

Figure 2.5 A control chart generated in 1-2-3 to monitor the performance of an instrument. A standard check sample is analyzed by the instrument on a regular schedule (each hour, day, week). The control chart displayed shows the results performed on a daily schedule. The results from the current analysis, the X symbol, are compared with the historical results. Three lines are plotted for the average, the average plus one standard deviation and the average minus one standard deviation. If the result is not within a specified range, the instrument should be checked to insure it is operating properly.

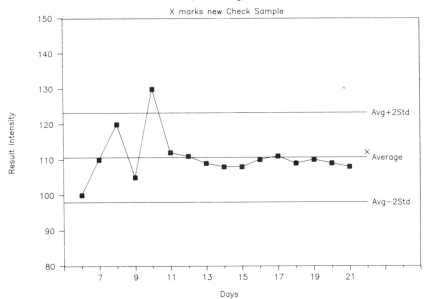

Semi-Automatic

The second category has automatic data acquisition but with data analysis performed via user entered commands. An example of this application category is the monitoring of the performance of a chromatograph. A check sample can be analyzed and the results of the sample stored in 1-2-3 to compare with the current results. The check sample could be automatically run each day (hour, week, month) and the data automatically acquired and placed into 1-2-3 and then combined with historical data to create control charts like the one shown in Figure 2.5. Data analysis options could be programmed by using a 1-2-3 macro with menu selections.

The third category is command driven data acquisition and data analysis. This is the simplest form of data acquisition and analysis since the user commands the system to perform both data acquisition and data analysis. If an error is observed during data acquisition the user can re-acquire the data. If an error occurs during data analysis, the user can start the data analysis procedure again.

Real-Time and Automatic Data Acquisition with 1-2-3

Both real-time and automatic data acquisition applications require special 1-2-3 add-in programs like Lotus Measure to automatically bring data into 1-2-3. Lotus Measure can automatically enter data into 1-2-3 through either RS-232 or IEEE-488 communication ports. Lotus Measure can also acquire data from the MetraByte DAS-16 data acqui-

Lotus in the Lab

sition card. This card is available from MetraByte Corporation whose location and telephone number is given at the end of this chapter.

If all of your data is communicated via an RS-232 port you could also use a communications program like Procomm or the communications sheet in Lotus Symphony to acquire data in real-time or automatically. Procomm is available in shareware version from many bulletin boards and users groups. See the end of this chapter for a contact.

Acquiring Data via Communications Ports

Most instruments can send all or some of their data to your PC through a data communications port. This section will provide step-by-step instructions on how to acquire data via RS-232 and IEEE-488 ports.

In many cases the data you would like to analyze in 1-2-3 is produced by an instrument with a communications port. Most modern instruments are controlled by a microprocessor based computer which can send data to a PC if the capability is implemented. It may be possible to acquire the data from the instrument and load it into 1-2-3 as one unit and thus completely skip manual data entry. Acquiring the data on your PC will also provide the data in a computer readable form which can be saved for documentation purposes or sent to other computers.

Command Driven Data Capture

This section will focus on simple command driven data acquisition into a disk file. We will not attempt to cover real-time data acquisition and analysis. Even if you need real-time or automatic data acquisition and analysis it is a good strategy to first implement a command driven application. Then plan a second development step to add in the real-time or automatic data acquisition and analysis capabilities. By implementing a command driven application first you will encounter and solve any problems early in the project. It is much easier to test systems which are not real-time because errors can be more easily uncovered, so your development will proceed much faster. Then when you start working on the real-time steps, you will know the majority of your system is sound.

Command driven data acquisition requires two major steps. First the data is acquired from the instrument into a disk file. The second step is loading the data into 1-2-3 using the /File Import and possibly the /Data Parse commands. Loading data into 1-2-3 will be covered in Chapter 3.

Data Communication Basics

To understand how computers and instruments communicate, we must first learn some basic terms and concepts. Signals between computers and between computer components are carried through metallic carriers upon which voltages may be applied and currents made to flow. These metallic carriers can be the equivalent of single wires or a combined set of wires called a bus. The simplest signal which can travel along such a conductor is the presence or absence of voltage or current flow. This is a binary signal since it can assume only two states: present or absent, ON or OFF, 1 or 0. The basic unit of such a binary signal system is a Binary digIT (BIT) which can have one of two values, 0 or 1. Binary signals are the primary means of communication used by computer systems because the circuitry required to generate and detect the presence or absence of a signal is much simpler than circuits which can measure the amount of signal present.

Using a voltage signal, the voltage is either a specified number of volts or zero volts. Voltage is measured with reference to a zero point, called ground, which is a heavy conductor interconnecting all components in the computer system. As long as there is agreement on the signal levels the circuits must send and receive, we have a hardware system for sending and receiving signals. A common binary signal level used within computers is Transistor-Transistor Logic or TTL. TTL voltage regions define a valid binary signal. The regions are:

> 2 to 5 volts — high region
> 0.8 to 2 volts — undefined region
> 0 to 0.8 volts — low region

We will see later in this chapter that RS-232 communications uses a different set of voltage regions to define a binary signal.

Character Codes

A binary language is fine for internal computer communication, but how can information be represented in a form we can understand? The answer is to use more than one signal line or bit. If eight bits are combined, we can represent 2 raised to the eighth power or 256 values. A combination of eight bits is called a byte. Then what is needed is a standard which defines the combination of eight 0's and 1's which represents each symbol. That is why ASCII (American Standard Code for Information Interchange and pronounced "ask-ey") was created.

Official ASCII is actually a seven-bit code representing 128 characters. This is enough to represent all the characters in the alphabet, both uppercase and lowercase, all the numbers 0-9, the punctuation marks found on a typewriter and special non-printing control codes like carriage return and line feed. The ASCII codes for the capital letters A-G are shown in Figure 2.6. A complete ASCII table used by the IBM PC can be found in Appendix A. By including the eighth bit, 128 more characters can be defined. On the IBM PC these additional codes are used to define special printable symbols and foreign alphabet symbols.

Figure 2.6 ASCII binary character codes for the capital letters A-G. The decimal, hexidecimal and octal representations are also shown for reference. When seven bits are sent, the left most bit is not sent.

```
A   0100  0001
B   0100  0010
C   0100  0011
D   0100  0100
E   0100  0101
F   0100  0110
G   0100  0111
```

Serial vs. Parallel Communication

There are two ways eight bits can be communicated. One way is to use eight wires each carrying a single bit of the eight bits. This is called parallel communication. Three common parallel communication standards are: parallel printer interface (sometimes called a Centronix interface) used by many PC printers, IEEE-488 used by scientific instruments and SCSI (Small Computer System Interface) used for many disk drives and communications equipment. These are all standards for communicating over relatively short distances. If communications is over a long distance, the cost of running several wires becomes prohibitive. In those cases, serial communications using a single wire is much less expensive. Serial communication sends the sequence of 0's and 1's separated by time. RS-232 is a popular serial communications standard.

Now that we understand a few more basic communication terms and concepts, we can move on to getting instrument data onto our disks.

Acquiring Data in a Disk File via RS-232

Many instruments and most computers have an RS-232 port or offer the port as a hardware option. RS-232 is not the easiest communications port to utilize because it has many parameters and configurations involving both hardware and software which must be matched for communication. But, with a little patience you will be able to acquire the data you need. It is easiest to understand how RS-232 operates by first describing the target application for the standard. The RS-232 standard (RS stands for Recommended Standard) was developed by the Electronic Industries Association (EIA) in the 1960's to standardize communication between computers and perpherial devices. Some books and manuals refer to this standard as RS-232C which was the last revision of the standard. In this book the C will be left off and the standard will be called just RS-232.

Figure 2.7 RS-232
communication was de-
veloped so a terminal us-
ing a modem could com-
municate with a remote
computer using telephone
lines.

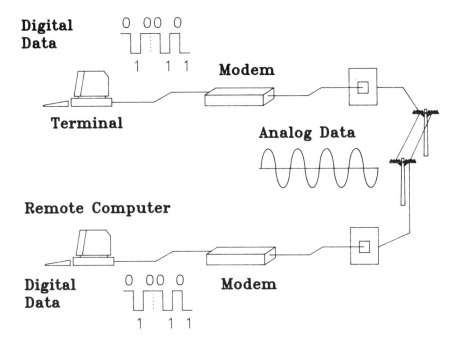

RS-232 was developed to allow a remote computer users to communicate with a computer with a terminal using regular telephone lines as shown in Figure 2.7. A key perpherial in this scheme is a modem (MOdulator DEModulator), a device which can convert digital values into audible sounds which can be carried over regular telephone lines. Two modems are needed for terminals and computers to communicate. One modem converts the digital codes sent by the terminal into audible sounds. The sounds are carried over telephone lines to the second modem. The second modem converts the sounds back into digital data. This process is reversed to send data from the computer back to the terminal.

Since only a single wire can be used, RS-232 is a serial standard. A positive voltage between 3 and 25 volts is used to represent a logic 0 level. A negative voltage between –3 and –25 volts represents a logic 1 level. (Note: These levels are used only for the data lines on pins 2 and 3; they are negative-true. All other signal in RS-232 are positive-true, meaning a positive voltage represents a logic 1 and a negative voltage represents a logic 0.)

Bits of a character are separated by time, thus a waveform is produced on the data line when a character is transmitted. Such a waveform for the transmission of the ASCII character "G" is shown in Figure 2.8. The

ASCII code for "G" is 1000111 in binary, and its least significant bit (the right most bit as written) is transmitted first.

Sending Data

RS-232 uses an asychronous method of data transmission. This means the sending and receiving computers do not have to be synchronized on the same time base. Synchronous data transmission is different and will not be discussed since it is rarely used with PCs. In asychronous data transmission a start bit is always sent to mark the beginning of a character. Following the start bit the data bits are sent. Each bit is held on the the data line for a precise amount of time call a bit time. The receiver times the incoming signal and samples each bit near the center of each bit time. The sender and receiver must agree on the length of time a bit will be held on the data line otherwise the communication will be incorrect.

The bit time determines the maximum rate characters can be transmitted. Common rates are 300, 1200, 2400 and 9600 bits per second (bps or baud, Bits AUDible). Following the last data bit a parity bit may be sent. This bit is used for error detection. A noise pulse or other problem could affect the data line causing one bit to be misread. If the transmitter keeps track of the number of 1's in the character being transmitted, it can set the parity bit so the total number of 1's is always even (for even parity) or odd (for odd parity). The receiver can also keep track of the number of 1's and determine if the transmission was received correctly.

Figure 2.8 Data transmitted to send the letter capital G. At the beginning of the transmission the signal line is held in the idle state (-3 to -25 volts). When data transmission starts, a start bit is sent which set the signal voltage between +3 and +25 volts. Then the seven data bits representing the the capital letter G are sent. The parity bit is then sent followed by the stop bit.

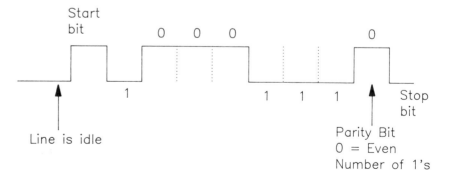

Capital G Binary 1000111

7 Data bits with Even Parity

Finally, stop bits are transmitted. These are not really bits but just a period of time to allow the components to prepare for the next character.

For a communication to be successful both the transmitting and receiving systems must have the same values for bits per second (bps or baud), number of data bits, parity bit (odd or even or not used), and the number of stop bits.

Controlling the Data Flow

Although RS-232 communications could be performed with as few as three wires, there are up to seven additional wires which are used to control the transmission of data. Unfortunately, even though these additional signal wires exist, they are normally not used for true data flow control. With RS-232 on a PC, data flow control (other than starting a transmission) and data communication error trapping must be done in software.

With the growth in popularity of computer bulletin board systems, many error checking communication protocols have been developed to control and trap errors using RS-232. Protocols like Kermit, XMODEM and YMODEM have been implemented for many computer types so that different computers can exchange ASCII data. The same protocol program must be run by both the sending and receiving computers. The sending computer, for example, sends 128 bytes followed by an error-checking value called a checksum. The checksum is calculated by adding the ASCII values of each character sent in the 128 bytes. The resulting value is then divided by 255 and the remainder is the checksum. The receiving computer performs the same calculation on the incoming 128 bytes. If the two values are the same, the next 128 bytes are signaled to be sent. If the values are not the same, the sender is signaled to send the 128 bytes again. Using such a protocol, data can be accurately communicated via RS-232.

Communication protocols are implemented in nearly all of the popular PC communication programs like Procomm, QMODEM and PCTalk. See the end of this chapter for a source for these programs. Unfortunately, very rarely do instruments which communicate via RS-232 have any communication protocols implemented.

The Physical Connection: DTE and DCE

Two types of devices were described in the original standard, DTE (Data Terminal Equipment) and DCE (Data communications Equipment). The DTE was the terminal while the DCE was the modem. The RS-232 port on most PCs are configured as DTE with the assigned pin configuration shown in Figure 2.9.

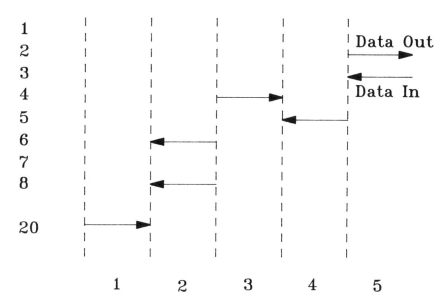

Figure 2.9 The pin configuration and control signal sequence used by a normal RS-232 port on a PC. The signals are voltages applied to the wires. A DC voltage of greater than +3 volts is ON while a voltage of less than -3 volts is OFF. 1) When the PC RS-232 port is ready to transmit or receive data, it raises the voltage on pin 20 to ON. 2) When the DCE device is ready to communicate it sets two lines ON: pin 6, DSR (Data Set Ready) and pin 8, DCD (Data Carrier Detect). 3) When the PC senses pins 6 and 8 ON, it then turns on pin 4, RTS (Request to Send). RTS signals the DCE that the PC has data ready to send. 4) When the DCE is ready to receive data it activates pin 5 to ON, CTS (Clear To Send). 5) Once the PC senses pin 5 ON it starts transmitting data on pin 2, Txd (Transmit Data). Similarly, data can be received on pin 3, Rxd (Receive Data).

If your target instrument is wired as a DCE you normally will have no problems as long as the vendor has implemented all of the necessary control lines. Problems will arise if the vendor does not provide all of the control lines, or if the sequence of when the control lines are activated is not correct for use with the PC. If the control lines are not present or they are not activated in the right sequence you will have to ignor the control signals from the instrument and wire an interface to provide the signal from another source. When this is done, you risk losing data since you will be receiving an uncontrolled transmission. It is like opening up a water valve with no way to signal a closure if the flow is too fast. If you have to use uncontrolled transmissions, it is best to start with a slow baud rate like 300.

The most common problem occurs when your target instrument is also configured as a DTE. Now you must wire a DTE-to-DTE interface which will fool each side of the communications link into believing it is sending and receiving data via a DCE. The input voltages on both ends must be present and occur in the correct sequences. The sequence of when the control signals occur is dependent on how the RS-232 port is implemented on the instrument.

Steps to Communicate via RS-232

All applications on your PC require two major parts, hardware and software. Normally you focus on the software aspects of an application but in this case you will be faced with a hardware task. Follow these steps to acquire data from your instrument using RS-232.

Preliminary Steps If you fail to complete any of these steps you should still go on to "Making the Hardware Connection" to definitely verify your findings.

Step 1: Find another user who has already acquired data from the instrument.

Make sure there is not already a method for you to acquire the data. A few telephone calls to the vendor or to other users could save you hours of work. With the thousands of PCs being put into use each day, you may be pleasantly surprised that someone has already performed and perfected the necessary steps to acquire the data you need. If you find someone, get the information in detail and then begin the implementation starting in the "Making the Hardware Connection" section.

Step 2: Make sure the instrument has an "active" communications port.

Active is emphasized because in many cases there may be a connector on the instrument or reference to a communications port in the operators manual or sales literature but the port is not operational or could be an option you failed to purchase. This step sometimes requires some real detective work. Normally a phone call to the instrument vendor should get you an answer. If you can not get an answer from the vendor, continue on to the next steps. You will get the real answer to this question by trying to actually make the connection.

Step 3: Make sure the communications port will send the data you want to acquire.

Even if the instrument has an active communications port, the data you need must be sent through that port. This is really dependent on the vendor. Most vendors will send the data which would normally be printed on a printer to the communications port. But you must be sure the the data you would like to acquire is sent through the port. Some vendors will disappoint you and will fail to send out the data you need. Again, if you can not get an answer you should still proceed on to the next steps.

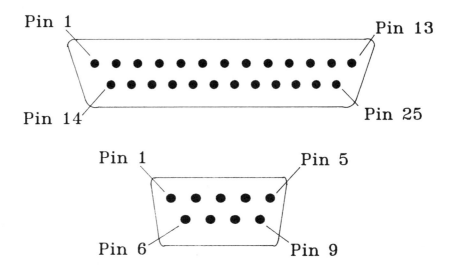

Step 4: Make sure your PC has an RS-232 port.

RS-232 ports are now quite common as part of multifunction interfacing cards. RS-232 ports are now standard on IBM PC/AT systems. Many PC clones have RS-232 ports as part of their standard configuration. You can also purchase single function RS-232 port interface cards for about $50.

Making the Hardware Connection

Step 5: Identify the RS-232 port on your PC, identify the type of connector and its gender.

Until recently, the standard RS-232 connector found on PCs and PC-XTs was a 25-pin male connector, as shown in Figure 2.10a. With the introduction of the IBM PC/AT the standard is a 9-pin male connector as shown in Figure 2.10b. A male connector has visible pins, while a female connector has recessed connections which look like holes.

Step 6: Identify the RS-232 port on the instrument, identify the type of connector and its gender.

The RS-232 port on the instrument could have one of many configurations. Normally, it will have a 25-pin connector. If the instrument does not have a 25-pin (or 9-pin) connector, then you will have to build an interfacing connector to make the hardware connection. Connectors are available at most stores which carry computer or electronic components like Radio Shack Stores. You may need to get some assistance from someone with some wiring and soldering experience.

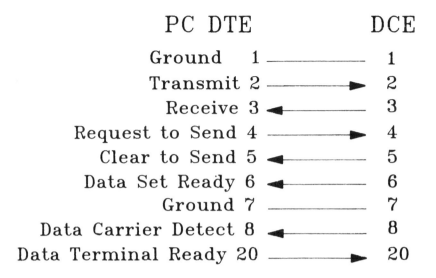

Step 7: Configure a cable to connect the two RS-232 ports.

Identify which pins are being used to transmit and control data transmission on the instrument. The simplest connection between two RS-232 devices is shown in Figure 2.11. You should be able to find this information in the instrument's operators manual in a section describing the communications port. Again, you may need to contact the vendor to get this information. The description may be very terse like "the RS-232 communications port is wired as a standard DTE (or DCE) device". Make a diagram like that shown in Figure 2.12a and fill in which lines are being used. Now you must connect the data and control lines.

Your cable configuration must be tested to see if it will work. These tests are best performed using a piece of diagnostic equipment called a break-out box. A break-out box is an indespensible piece of equipment for making the initial hardware connections and for trouble shooting connections. A large assortment of these break-out boxes are available at computer/electronics stores and through mail order catalogs. Prices range from $49 to $219 dollars. A test box normally has two 25-pin connectors (9-pin versions are also available) so it can be easily connected into the RS-232 line between the two devices to be tested Figure 2.13. The break-out box may also have LED's (light emitting diodes) to monitor all lines which might be active. An active line will illuminate the LED. The box will also have either switches or wires so you can connect specific incoming pins with outgoing pins. You can thus create the connection necessary and test the configuration.

Figure 2.12 Identifying the data and control lines. Some lines will need to be crossed and others jumpered. Figure 2.12a shows a blank worksheet to fill in for your instrument. Identify the transmitting and receiving lines along with the control lines your instrument can provide. If your instrument transmits data from pin 2 you will have to cross lines 2 and 3 as shown in Figure 2.12b. If your instrument does not provide any signals then you will have to jumper the signal out on pin 20 to go to pins 5, 6 and 8 as shown in Figure 2.12c. Draw lines and arrows showing the line connections for your particular instrument.

PC DTE

Ground   1 ——————
Transmit   2 ——————▶
Receive   3 ◀——————
Request to Send   4 ——————▶
Clear to Send   5 ◀——————
Data Set Ready   6 ◀——————
Ground   7 ——————
Data Carrier Detect   8 ◀——————
Data Terminal Ready 20 ——————▶

(a)

2 ＼   ／ 2

3 ／   ＼ 3

(b)

5 ◀
6 ◀
8 ◀
20 ——▶

(c)

Figure 2.13 Typical break-out box which allows you to experiment with the various wiring configurations without having to build different cables and connectors. Most break-out boxes also include LED (light emitting diodes) displays to monitor the activity on the various lines.

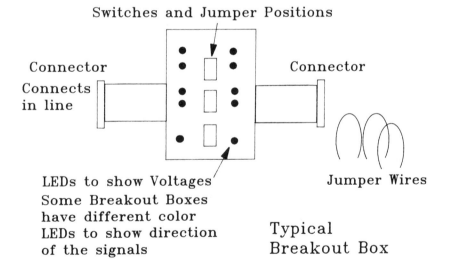

Switches and Jumper Positions

Connector
Connects
in line

Connector

LEDs to show Voltages
Some Breakout Boxes
have different color
LEDs to show direction
of the signals

Jumper Wires

Typical
Breakout Box

Make sure the cable you purchase has enough lines (wires) to carry all of the signals needed to make the connection. Cable is available with 4-, 9-, 15- or all 25-lines. The RS-232 port on a standard PC will utilize up to 9-lines. You should also check the cable continuity, meaning a

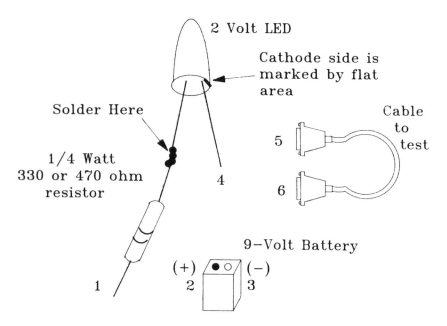

Figure 2.14 Line continuity tester. This simple tester can be used to insure an electrical voltage can be transmitted through a cable and its connectors (continuity). It can also detect when the wires in a cable are crossed. The tester is made of parts you can purchase at any electronics store: a common 9 volt battery, a 2 volt LED (light emitting diode) of any color, three sets of alligator clips (alligator clips on each end of a short piece of wire), wire, electrical tape or shrink wrap and a 1/4-Watt 330 or 470 ohm resistor. Assemble the parts as shown above. You have built a simple circuit. When you apply a voltage by attaching the anode side of the device to the + side of the battery and the cathode side to the - side of the battery, the LED will illuminate. Test your work to insure the LED lights To test a cable, attach the alligator clips, battery, testing device and cable as shown. If the cable and connectors are sound, the LED will light.

voltage signal will flow continuously from the pin on one end of the cable to the pin on the other end of the cable. This can be done by making the simple tester in Figure 2.14 or by using a more elegant voltmeter. Some cables will be purposely wired to cross-over signals. For example, pin 2 on one end will be connected to pin 3 on the other end. Be sure you know how all cables you are using are wired. You can use the cable tester in Figure 2.14 or a voltmeter to also test the connectivity of the cable wiring.

Null Modems

Some common DTE-DTE connections can be solved by using a null modem. A null modem is a device or cable with 25-pin connectors on each end. The wires connecting the pins on each end of the null modem are redirected to send the signal for example from pin 3 on one end to pin 2 on the other end. To redirect control signals, some pins are even jumpered. All null modems are not wired the same. Figure 2.15 shows four different

Null Modem A   Null Modem B

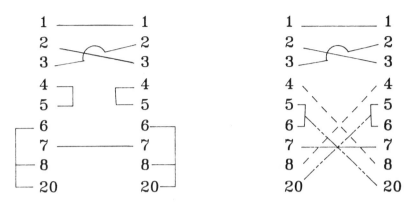

Figure 2.15 Null modem wiring configurations from two vendors. All null modems are not the same! Notice the variety of different wiring configurations which are offered by these two vendors. These represent solutions to RS-232 interfacing problems for some PC users and are an example of the wide array of possible combinations of solutions you may need to implement. The control signals must be present and the sequence of when they are activated must be correct, otherwise no communications or uncontrolled communications will occur.

```
10  CLOSE
15  INPUT "Filename for Storage: ",F$
16  OPEN F$ FOR OUTPUT AS #1
20  A$=""
30  OPEN "COM1:1200,E,7,1,CS,DS,CD" AS #2
40  '
50  'COM1 opened with 1200 baud, Even parity, 7 data bits, 1 stop bit.
60  'The CS, DS and CD tell the port to not check the status on the control
70  'lines, Clear to Send (CS), Data Set Ready (DS) and Carrier Detect (CD)
80  'Enter the values the instrument is using here .......
90  '
100 INPUT #2,A$: 'Input data from the RS-232 port
110 PRINT A$: 'print the data on the screen
120 WRITE #1,A$: 'Write the data to the disk
130 GOTO 100:' Do it again
```

Figure 2.16 Simple BASIC program to receive data through the RS-232 port. The communication parameters you must set are dependent on the parameters used by the instrument. The parameters you must set are the communications line you are using, baud rate, parity, the number of data bits, and stop bits. These values are set in the OPEN "COM ..." statement in line 30. This program does not utilize any of the control signals which can be controlled when opening the communications channel. Read more about control in your BASIC Reference manual under the OPEN COM statement

null modem wiring diagrams. Null modems can also be used to solve the connector gender problems since they can be purchased with either male-male, female-female or male-female connectors on each end.

Step 8: Send some Data from the Instrument to your PC.

Once you have your hardware connected, you can test it by sending some data. A simple BASIC program to receive data is shown in Figure 2.16. Type in this program and enter the proper communication parameters for the data transmission. The values for these parameters should be setable on the instrument or if not, the vendor should provide the parameter values.

The BASIC program will print on the screen the text flowing over to your PC. The data will also be stored in a disk file just as it is shown on the screen. Once the data is acquired, you can stop the program by pressing Ctrl-Break. Close the disk file by entering the command CLOSE and press the enter key.

If you have any experience with using a communications program like PC-Talk, QMODEM, Procomm, Smartcom, CrossTalk, SmartTerm etc., you can run any of those programs to set the communication parameters and acquire the data. Communication programs will assume there is a modem sending the data so your instrument or the cable you have configured must have the proper signal and control lines active for the communication program to work. Use the communication program's feature to download data into a disk file to acquire the data.

Step 9: Refining Data Capture

You now have acquired some data. Congratulate yourself. The small BASIC program is the primitive core to a program you will need to acquire data. You can add many more features to this core program such as the ability to perform two-way communications to trigger the instrument for data, control the transmission of data from the instrument by using one or more control lines, or graciously end data acquisition when no more data is available. Your acquired data can be loaded into 1-2-3 using the commands and techniques in Chapter 3.

If you are unsuccessful at this point in acquiring some data, do not be discouraged. Whole books have been written about interfacing via RS-232 because of the many potential problems. If you can not make the communications link by following these steps you should consult one of the following books and keep trying: *The RS-232 Solution,* by Joe Campbell, SYBEX Books, Berkeley, CA; *RS-232 Made Easy: Connecting Computer, Printers, Terminals and Modems,* by Martin D. Seyer, Prentice-Hall.

Capturing Data via IEEE-488

In the early years of computing, computer designers tried to incorporate the latest components and features into both computers and peripherals. As each new generation of hardware was developed, new communication interfaces were created to take advantage of new and faster components. The result was a multitude of interfaces, few of them compatible with any of the others.

The interface incompatibility problem was recognized by instrumentation developers and in 1975 an interfacing standard was agreed upon and published by the Institute of Electrical and Electronics Engineers (IEEE). It was the first standard for interfacing computers, peripherals and instrumentation. The first version, IEEE Standard Digital Interface for Programmable Instrumentation (IEEE-STD-488-1975) was slightly revised in 1978 and is now IEEE-STD-488-1978 commonly called just IEEE-488.

The standard defines a general purpose interface designed to connect instrumentation systems requiring limited-distance communications. The intent of the standard was to fix as many variables of an interface as possible while still allowing flexibility for implementation. This has allowed the interface to be compatible in most cases, even though new, faster and less costly components are now being used. The interface is a parallel input/output bus with specific signal levels and signal timing. Different devices can connect to the bus and communicate with each other by sending electronic messages to each other.

The standard interface allows you to build an instrument system by simply connecting various instrumentation units together with IEEE-488 interfaces. Three types of devices exist on the bus: controllers, talkers and listeners. At any specific time, there can be only one controller. There can be up to 16 other devices which can be talkers or listeners or both. The controller is like the chairperson of a meeting. It controls communications on the bus by allowing only one talker at a time to supply the bus with information.

Usually your PC will be the bus controller and also have talking and listening capabilities. To send and receive data using an IEEE-488 interface, you will need an interface card. These cards are available from a number of vendors. One vendor is Ziatech Corporation. Their address and telephone number will be printed at the end of this chapter. Software to perform all basic communications using the interface bus will be included with the interface card. The software is program code you can add to an existing program or build a program around.

If you have an IEEE-488 device you wish to acquire data from, follow these steps.

Preliminary Steps If you fail to complete any of these steps you should still go on to "Making the IEEE-488 Hardware Connection" to definitely verify your findings.

Step 1: Find someone who has already acquired data from the instrument via IEEE-488.

Make sure there is not already a method for you to acquire the data. A few telephone calls to the instrument vendor or to other users could save you some work. With the thousands of PCs being put into use each day, you may be pleasantly surprised that someone has already performed and perfected the necessary steps to acquire the data you need. If you get lucky and find someone, get the information in detail and then begin the implementation at step 5.

Step 2: Make sure the instrument has an "active" IEEE-488 port.

Active is emphasized because in many cases there may be a connector on the instrument or reference to a communications port in the operators manual or sales literature, but the port is not operational or could be an option you failed to purchase. This step sometimes requires some real detective work. Normally a phone call to the instrument vendor should get you an answer. If you can not get an answer, continue on to the next steps. You will get the answer to this question by trying to actually make the connection.

Step 3: Make sure the IEEE-488 port will send the data you want to acquire.

Even if the instrument has an active IEEE-488 port, the data you need must be sent through that port. This is really dependent on the vendor. Most vendors will send the data which would normally be printed on a printer to the communications port. But you must be sure the the data you would like to acquire is sent through the port. Some vendors will disappoint you and will fail to send out the data you need. Again, if you can not get a clear answer you should still proceed to step 5.

Step 4: Make sure your PC has an IEEE-488 port.

IEEE-488 interface cards are now widely available from many different vendors. Read the instructions on how to install the interface card and make sure the card is operating properly. Most vendors provide diagnostic software to insure the interface card is operating properly.

Making the IEEE-488 Hardware Connection

Step 5: Connecting two or more IEEE-488 components is very easy. The interface connectors have two sides which allow cables to be stacked. This configuration is also called daisy-chains since multiple lines can radiate out from a central connection like the flower petals of a daisy. The connectors can only be joined in one way, so it is not possible to make a misconnection.

Step 6: Send some Data from the Instrument to your PC.

Once you have your hardware connected, you can test it by sending some data. If you are unable to acquire the data, check your cables and connections again. Then make sure the software is operating correctly. You can complete the data acquisition process by storing the data on disk.

IEEE-488 data acquisition is normally straight forward. This is because the standard is much more detailed and most implementations meet all aspects of the standard.

Now we will look into acquiring data directly from analog signal sources. This requires a data acquisition interface.

Data Acquisition Basics

Data acquisition and control are cornerstone applications for PCs and computers in general. As scientists, we make observations to better understand the world around us. Computers can help us make these observations by performing otherwise boring redundant tasks such as monitoring the temperature for a reaction over a 24-hour period. Computers can acquire data on events which take fractions of a second, allowing us to record an event which otherwise would go undetected with human senses. Finally a computer can store data, giving us the ability to display or manipulate the results of the observation allowing us to use the computer for data analysis and reporting. By using a computer we also expect better data quality since there should be no possibility of human error unless it is part of the initial experimental design.

Data acquisition and control are applications which should interest nearly every scientific user of a PC. Modern data acquisition systems can be cost-effectively created using a PC as an integral component. Data acquisition products for the PC bring together a number of components into an integrated system allowing the user to utilize the technology without being intimately familiar with all the internal details. Even with these advanced products, you will need to master some basic data acquisition and control concepts so you can correctly analyze your data and plan experiments.

Digitizing an Analog World

We perceive the world around us as continuous or analog. Signals to preceive light, sound, touch and smell are all sent continuously to our brain. Computers unfortunately are able to manipulate only discrete or digital data. Thus a major part of data acquisition is to take analog observations or stimuli and convert them into digital or discrete values. In essence a snapshot of the analog observation or stimulus is made so a digital computer can be utilized. The conversion process, the step to convert the observation or stimulus into a digital value, dictates the quality of the resulting data. If the conversion process is poorly performed, the resulting digital data can be useless and renders any deductions made from the data erroneous.

Acquiring data is much like taking pictures with a camera, a number of problems can occur which will cause poor results. If the camera lens, shutter speed or aperture settings are not correct there will not be enough light to activate the film. In addition, even if the film is activated nicely, the wrong chemicals could be used to develop the film and cause poor results. All parts of a data acquisition system must be designed and implemented correctly to obtain good results.

There will always be some amount of error involved in any data conversion. Usually the more accurate you need to make the conversion, the more it will cost in terms of price for electronic components. It is up to us to know the amount of potential error and plan our experiments accordingly. A certain amount of error can usually be tolerated in a measurement without effecting the end result. By understanding the tolerance in a measurement, the time or costs required for implementation and operation of the measurement system can be optimized.

The results of such an analysis for cost or time savings can be found in your own car. The control panels on most cars today have very few analog meters. Usually only the gasoline gauge is an analog meter. To monitor the engine temperature or battery charge simple on/off lights are present which are illuminated when the temperature is too high or your battery is not holding a charge. These simple lights can warn you of a problem. Meanwhile, the manufacturer has saved the difference in cost between a meter and a light emitting diode.

Where Did This Data Come From?

The block diagram in Figure 2.17 can be used to describe how most data is acquired into a computer. The observation or stimulus occurs and is sensed by a transducer of some type. A transducer is a device or circuit that converts physical parameters such as position, temperature, pH, light intensity, force, flow, acceleration, velocity etc. into an electrical signal such as voltage, current, charge, resistance or capacitance for purposes of measurement or control. The actual form a specific transducer takes depends on the intended function and manufacturer's preference.

In our chromatography example, the major transducer is the detector. It converts the amounts of a compound into a voltage. There are many types of chromatography detectors. Some measure the minute change in temperature (thermal conductivity), or the amount of charged particles produced when the sample is burned (flame ionization) as a compound flows through the detector. Others measure the amount of ultraviolet light or the specific wavelength of light a compound absorbs.

Complete system manufacturers build products which include all of the components in one integrated unit. Most versions of these systems have the ability to share their data with a PC. Usually this is done via a common communications interface such as RS-232 or IEEE-488.

Figure 2.17 A stimulus is detected by a transducer which converts the stimulus into an output voltage E1. The voltage E1 is then amplified to E2 to be compatible with the input requirements of an analog-to-digital (A/D) converter. The A/D converts the analog voltage into a digital value which can be stored and utilized by the computer. The output voltage from the transducer is then normally amplified to meet the voltage range of the analog-to-digital (A/D) converter. The A/D converters transform the analog voltage into its digital representation and provide the digital value to the computer. The components to assemble a data acquisition system and complete systems to perform these tasks are available from many manufacturers.

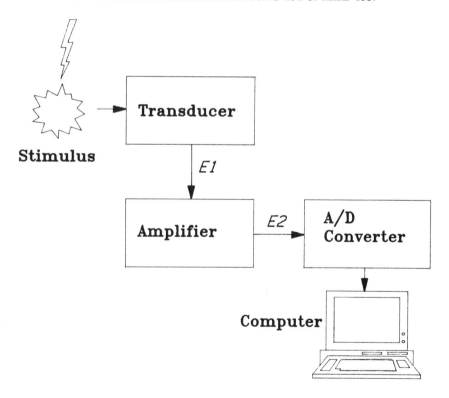

Figure 2.18 Data ac-
quisition interface board
with screw connector
module.

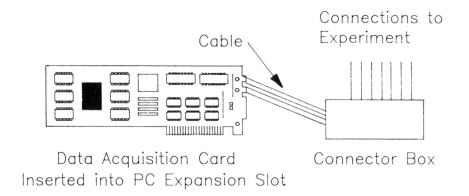

Cable

Connections to
Experiment

Data Acquisition Card
Inserted into PC Expansion Slot

Connector Box

Component manufacturers build products which will for example contain a transducer with an amplifier. The user can then select a separate
A/D converter to interface the transducer with the computer. A list of
some manufacturers of both complete systems and components is printed
at the end of this chapter.

PC Based Data Acquisition Systems

PC based data acquisition systems are usually assembled by selecting
components which best meet your needs. These components are available in many forms. Usually the transducer is selected or provided as
part of an instrument. The output voltage signal is then acquired by an
A/D converter. There are two types of configurations for the A/D converter and computer interface: plug-in boards and external stand-alone
units.

Plug-in Boards

A plug-in board fits right into a PC expansion slot. The board is pow-
ered by the PC power supply and uses the input/output bus of the PC to
communicate with the other computer components. Usually these plug-
in boards have the ability to monitor one or more analog signals, and
perform control via digital input/output lines or a digital-to-analog
(D/A) converter. The analog signal lines from a transducer are attached
to the board through a screw connector module Figure 2.18. The same
screw connector module is used for attaching digital input/output lines
or for making output analog connections. The board in the PC communicates with the screw connector box through a connecting cable.

The major advantages of plug-in board systems are their low cost and
flexibility for implementation. Major disadvantages are that a plug-in
board is made for only one computer type and the computer normally
must be near the experiment. For example, a plug-in data acquisition
card for an IBM PC will not work in an Apple II computer. Nor will a
card made for an IBM PC work in the new IBM PS/2 using the

Microchannel bus. The computer must also be close to the experiment because the signal lines from the transducers normally should not be allowed to span long distances. Electrical signals carried by wires over long distances can be affected by external magnetic and electrical fields which will cause the signals to weaken or be lost.

Stand-alone Systems

Stand-alone systems are contained in an external unit. This unit communicates with the PC via a standard communication interface such as RS-232 or IEEE-488. Using the communication interface, the PC can send the stand-alone system instructions on how to acquire the data. Then the unit acquires the data and sends the results over the communication interface to the PC.

The major advantage of a stand-alone system is that it can be used with any type of computer with the proper communication interface. The external unit can usually be placed in location remote from the computer since communication of digital data can span longer distances. The major disadvantages include higher costs and less flexibility in implementation.

Minimizing Conversion Errors The two components in a data acquisition system which have the most influence on conversion errors are the transducer and A/D converter.

The Transducer

Transducers must cause an electrical parameter to vary with the applied physical stimulus. To accomplish transduction the physical stimulus must cause a change in electrical resistance, capacitance, inductance or some combination of these to produce an electrical current, a voltage, a frequency or digital word. The resulting voltage, frequency or digital word should be unique for each permissible value of the applied stimulus.

The most common forms of transducers use changes in resistance to indicate changes in the physical stimulus. Various materials exist that will transduce through thermoresistance, photoresistance, piezoresistance (deformation), or simply the position of a potentiometer. How precisely the output of a transducer tracks the change in the physical stimulus is described using four terms: linearity, hysteresis, precision and frequency response.

Linearity and Hysteresis

Linearity is a measure of how well a transducer meets an ideal calibration line or predictable function. A calibration line or curve is constructed by plotting stimulus values versus the transducer response. An ideal transducer will always respond to a specific stimulus with the corresponding value on the calibration line or curve. Real transducers have good linearity over only a specific range of stimulus values.

A log-log plot of the response of a chromatograph flame ionization detector is shown in Figure 2.19. Note the response is almost perfectly linear over the range 1 to 1.00E+7. Linearity then drops off significantly. This is a common observation with real transducers.

Hysteresis is the difference in the readings a transducer generates for the same stimulus depending on whether the stimulus value is approached from below (ascending) or from above (descending). A good transducer will provide the same reading whether the value is approached from above or from below.

Precision and Frequency Response

The precision of a transducer describes the reproducibility of a measurement. If an identical stimulus is repeatedly applied to a transducer, an identical result should be produced. Precision is a measure of how well a given transducer meets this ideal condition. Precision is normally reported as a standard deviation or percent standard deviation in converted units.

Figure 2.19 Linearity range of a flame ionization detector.

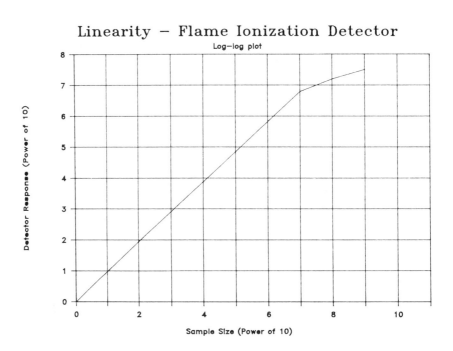

The frequency response of a transducer describes how fast the transducer will respond or track a change in the applied stimulus. If the frequency response of a transducer is too slow for an application, the transducer will be unable to detect the changes occuring in the stimulus and thus produce erroneous data. Similarly, if the frequency response is too fast, it may pass noise artifacts as real signals which will also produce poor data.

The Analog-to-Digital Converter

There are many types of analog-to-digital (A/D) converters. The various types and strategies used to select a converter type is beyond the scope of this discussion. You should refer to one or more of the books referenced at the end of this article for a discussion of A/D operation and types.

This discussion will focus on two important parameters of A/D conversion which strongly dictate the accuracy of the conversion. The parameters are the sampling rate and the resolution or dynamic range of the conversion.

Sampling Rate

Sampling rate is the number of A/D conversions performed per second. A good rule of thumb for selecting a sampling rate is provided by Nyquist's theorem. Nyquist's theorem says to set the sampling rate at two times the highest frequency event of interest. Figure 2.20 shows the results of sampling a sine curve at too slow a rate.

Sampling at too fast a rate can also produce poor results. Sampling at too fast a rate can cause any noise or random events to dominate the analysis of the data obscuring larger trends. Faster sampling rates also require larger amounts of data to be acquired and stored.

Resolution and Dynamic Range

Resolution and dynamic range dictate how many discrete values the analog signal level can be assigned. The number of discrete values is normally reported as the number of bits the digital value can be assigned. The most common resolutions found are 8-, 12- and 16-bits. The more resolution, the more expensive the A/D as is shown in Figure 2.21.

If the A/D does not have enough resolution, events in the data can be completely missed or poorly represented. An example of the results of acquiring a very small peak using various resolution A/D converters is shown in Figure 2.22.

Using a Data Acquisition Interface Card

A large number of vendors manufacture data acquisition interface cards for the IBM PC family of computers. A list of some of these vendors is printed at the end of this chapter. Data acquisition and control interface cards come in many combinations of capabilities. Most have 2-16

Figure 2.20 Examples of acquiring a sine curve at too slow a rate. The top graph show the sine wave acquired at 52 points over each cycle. The bottom graph is acquired at just 3 points per cycle and provides a very distorted view of the data.

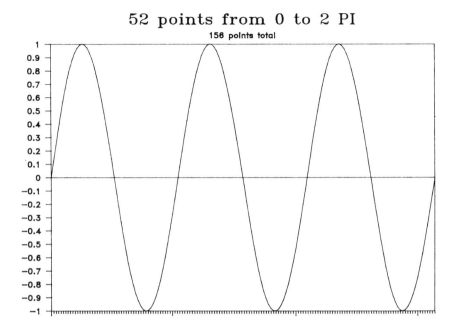

52 points from 0 to 2 PI
156 points total

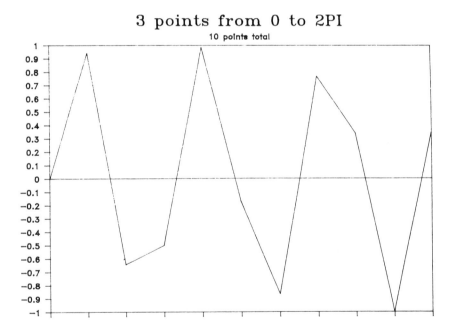

3 points from 0 to 2PI
10 points total

Figure 2.21 A/D con-
verter resolution and ex-
ample prices. These are
prices for the Analog
Connection interface
cards from Strawberry
Tree Computers. A termi-
nal box to make the con-
nections to the trans-
ducers is also required
and costs $295.

Number of Channels	Bits of Resolution	Price	Range of Values
8	12	$ 790	0- 4095
16	12	$ 990	0- 4095
8	16	$1290	0-65535
16	16	$1590	0-65535

channels of analog signal input and 2-16 digital input/output lines.
Digital input/output lines can be used as either input or output signal
lines. Output signals could be used to control a switch or relay. This
would require an external module containing the electromechanical
components for a switch or relay. Other interface cards also have digital-to-analog (D/A) output capability.

A good example of one of these data acquisition cards is from the
Analog Connection series of interface cards from Strawberry Tree
Computers. These cards offer a range of A/D conversion accuracy of 12-,
14- and 16-bits. Each card also has 16 digital input/output lines. Two
features of these boards make them attractive. First, they allow you to
connect a variety of different transducers using the same input connection including thermocouples, semiconductor temperature sensors, pressure transducers, strain gauges, flow sensors, potentiometers and 4 to 20
milliamp current loops.

Second, each board comes with a complete set of data acquisiton software. This software is menu driven and easy to use. The main program
is written in BASIC with assembly language routines for performing
fast data acquisition. If you need to write your own custom program, the
BASIC source code is provided. There is also a driver program which
can be controlled with six different high level languages including in-
terpreted BASIC, compiled BASIC, Pascal, C, assembly and through
the program ASYST. Acquired data can be displayed, graphed and
stored on disk. The program allows you to set alarm points, display
minimum, maximum and averages of all 8 or 16 channels depending on
the type of card you purchased.

Other data acquisition cards also provide some software. Usually the
software is in the form of assembly language routines which can be
called from high level languages like BASIC, Pascal, or C. To utilize
the interface card you must write programs which call the assembly
language routines, perform the calculations and create the displays you
wish to see.

Figure 2.22 Results of acquiring data using an 8-, 12-, 16- and 20-bit A/D converter. Notice the 8-bit converter was unable to digitize the signal because the highest part of the signal was below the lowest conversion value. The other converters can digitize the signal but the 12- and 16-bit A/Ds have rather large errors. Of course if you are not interested in very small signals or precise representation of the data, lower resolution A/Ds can be utilized.

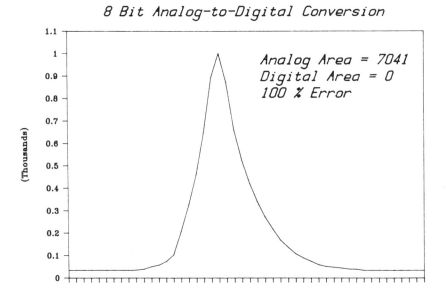

8 Bit Analog-to-Digital Conversion

Analog Area = 7041
Digital Area = 0
100 % Error

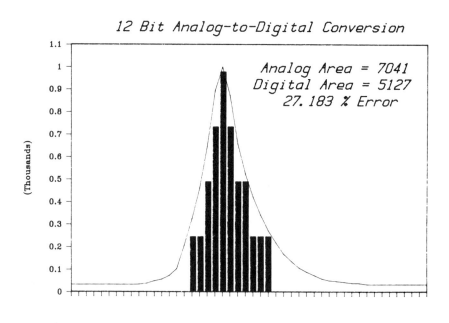

12 Bit Analog-to-Digital Conversion

Analog Area = 7041
Digital Area = 5127
27.183 % Error

(continued)

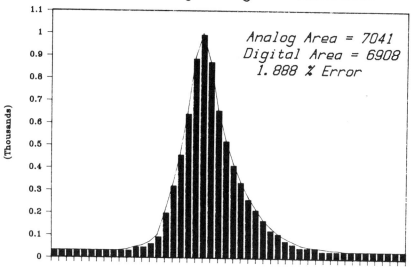

16 Bit Analog-to-Digital Conversion

Analog Area = 7041
Digital Area = 6908
1.888 % Error

(Thousands)

20 Bit Analog-to-Digital Conversion

Analog Area = 7041
Digital Area = 7054
.185 % Error

(Thousands)

Easy to use data acquisition software which supports many different data acquisition cards is also available. One of the most popular programs of this type is Labtech Notebook. Labtech Notebook can display data in real-time, perform curve fitting, perform fast fourier transforms and stores data on disk. This data can then be loaded into 1-2-3 for further data analysis and reporting.

Setting Up the Interface

Installing an interface card is quite easy. The card itself is installed in an open expansion slot. A ribbon cable is then attached to a connector on the card much like you attach a parallel printer cable. The other end of the ribbon cable is attached to the terminal box. The terminal box has a number of screw terminal connections for attaching the analog signal lines from the transducer.

After attaching the analog signal lines, you can load and run the data acquisition software. The menu driven software allows you to setup and acquire data, store the data on disk and display the data in various formats. The stored data can be easily loaded into 1-2-3 using the techniques described in Chapter 3.

Additional Reading

Here are three excellent books which will go into much more depth to cover computerized data acquisition.

Joseph J. Carr, *Digital Interfacing with an Analog World*, TAB Books Inc. (1978) 406 pages. This is an excellent book for learning the basics of computer interfacing.

John H. Moore, Christopher C. Davis and Michael A. Coplan, *Building Scientific Apparatus: A Practical Guide to Design and Construction*, Addison-Wesley (1983) 483 pages. This is an excellent hands-on book describing techniques for building scientific apparatus for experimentation. The book has over 140 pages dedicated to electronics and is an excellent resource for practical information for experiment design and implementation.

Kenneth L. Ratzlaff, *Introduction to Computer-Assisted Experimentation*, John Wiley & Sons (1987) 438 pages. This is an excellent book which provides the necessary background in electronics and electronic components to understand the many aspects of computerized data acquisition and control. The book takes you from preliminary concepts to complete system design and implementation.

Personal Computing Tools, Inc. 101 Church St. Unit 12, Los Gatos, CA 95032 (408) 395-6600. Supplies a number of data acquisition system components both interface cards and software. Specializes in hard to find PC hardware and software for scientific applications. Extends a risk-free 60 day trial period. Call for their latest catalog.

Burr-Brown Corp, P.O. Box 11400, Tucson, AZ 85734 (602) 746-1111.

MetraByte Corp, 440 Myles Standish Blvd., Taunton, MA 02780 (617) 880-3000.

Strawberry Tree Computers, Inc., 160 So. Wolfe Road, Sunnyvale, CA 94086 (408) 736-3083.

Data Translation, 100 Locke Dr., Marlboro, MA 01752 (617)481- 3700.

Laboratory Technologies, 255 Ballardvale Street, Wilmington, MA 01887 (617) 657-5400.

Keithly Data Acquisition and Control, 28775 Aurora Rd., Solon, OH 44139 (800) 552-1115.

Laboratory PC Users Group, 5989 Vista Loop, San Jose, CA 95124 (408) 723-0947. Distributes freeware/shareware versions of data communications programs like Procomm, QMODEM and PCTalk.

Ziatech Corp., 3433 Roberto Ct, San Luis Obispo, CA 93401 (805) 541-0488. IEEE-488 and other interfaces for IBM PCs.

3

Importing Data into 1-2-3

In this chapter we will investigate various techniques for importing disk data using the 1-2-3 File Import command, Data Parse command in 1-2-3 version 2, a BASIC program, a macro using the {READ} command and RAM Resident programs.

Importing Data Chunks with File Import

Lotus 1-2-3 provides easy to use methods for reading in chunks of data into worksheets. The simplest is to use the File Import command. This command will read text files off your disk and if the files are properly formatted, place the data into cells in the spreadsheet.

The data will be placed in the worksheet starting in the position where the cursor is placed when the File Import command is executed. Imported data is added to the current contents of the worksheet similar to the way the file combine command operates. Data in the files must be ASCII text. ASCII stands for the American Standard Code for Information Interchange. This is a code for representing text and special characters in a computer. There is an ASCII table in Appendix A of this book. Files using these codes are easily manipulated in your PC since most programs provide ways to read and write ASCII files.

Why Aren't all Files ASCII Text?

Disadvantages of ASCII text files are they require much more space on disk to store and they must be written and read sequentially. In BASIC programs, each ASCII text character requires a full byte (8 bits) to store. Numbers stored in ASCII files require one byte for each digit. For

Table 3.1 Number Representations in BASIC

TYPE	RANGE	ACCURACY
Integer	-32768 to 32767	0 decimal digits
Single Precision Floating Point	10E–38 to 10E+38	6 decimal digits
Double Precision Floating Point	10E–38 to 10E+38	16 decimal digits

example, the number 123 would require three bytes to store. If the same number is stored in binary format as an INTEGER it would require only two bytes, or as a SINGLE precision floating point number uses only 4 bytes and a DOUBLE precision floating point number requires 8 bytes. See Table 3.1 for the range and accuracy of these methods for representing numbers.

Do I Have an ASCII Text File?

The quickest way to see if the data you would like to import is in ASCII text format, use the DOS TYPE command to list the file. Enter the command:

```
TYPE filename.ext
```

where filename.ext is a typical result file you wish to import. If you see normal readable text on the screen, you have an ASCII file. If you see a series of weird characters and readable characters separated by blanks you do not have an ASCII text file and you will not be able to use the Import command. If you do not have an ASCII text file you should contact the instrument vendor and request to obtain the file type and structure. Most vendors will provide this information and many will even provide prewritten programs to convert the non-ASCII files structures into ASCII text. If your vendor provides only the file type and structure you will have to write your own conversion program.

Yes, I Have an ASCII text File

If you have an ASCII text file, you will be able to import the contents into 1-2-3. If you are using 1-2-3 Version 1A the file you wish to import must have the filename extension ".PRN". Files with other file extensions will not be imported with Version 1A, so you must RENAME or make copies of your files so they have the ".PRN" extension. 1-2-3 Version 2 allows you to import files with other extensions.

1-2-3 allows you to import data in two forms. The simplest form is straight TEXT. Each line of text in the imported file is placed in the spreadsheet as a LABEL. This form is not very useful especially if you are trying to import data values. But unless the file you are Importing is formated correctly, this is the only way you will be able to enter data.

Importing Values and Labels

To import both Values and Labels, the file to be imported must have the following structure: each numeric value must be separated by commas and labels must be delimitated by quotation ("....") marks. Each numeric value will then be placed into its own cell as numbers while text beginning and ending with quotes will be placed in their own cells as labels. Whenever a new line is found in the imported file, 1-2-3 moves down to the next row, moves back to the starting column and continues data importing. For example, a file with the name ABC.PRN has the following contents:

```
"Compound","Ret.Time","Area","Amount"
"METHANE",3.45,87654,45.89
"ETHANE",4.33,98765,98.54
"PROPANE",5.66,34567,32.87
```

In 1-2-3, with the cursor in cell A1, enter the command:

```
/File Import Numbers
```

then select the file ABC from the filename menu. The first line of labels will be placed in cells in the first row, A1, B1, C1 and D1. For the following lines, METHANE is placed in cell A2, the value 3.45 is in cell B2, and the values 87654 and 45.89 are in cells C2 and D2 respectively. Cell A3 contains the label ETHANE and the values for ethane are placed as values in cells B3, C3 and D3. Similarly, the values for propane are placed in row 4 as shown in Figure 3.1.

My Files do not Have the Right Format?

If your files do not have the right format to import numeric values correctly, numbers separated by commas and labels deliminated by quote marks, you will have to do a little more work. Depending on which version of 1-2-3 you are using you will have to do one of the following. Using 1-2-3 Version 2

Figure 3.1 ASCII text is imported into the 1-2-3 worksheet.

	A	B	C	D	E
1	Compound	Ret.Time	Area	Amount	
2	METHANE	3.45	87654	45.89	
3	ETHANE	4.33	98765	98.54	
4	PROPANE	5.66	34567	32.87	
5					

If you are using 1-2-3 Version 2, you will be able to use the Data Parse commands to separate the numbers and labels and place them into different cells on the spreadsheet.

Data Parse

Data Parse is a new set of commands which can separate a long single label residing in one cell, into a number of labels and numeric values contained in many cells. If your data is in a typical text file, one which would look nice printed on a printer or displayed on your screen, then it will have none of the necessary formatting as we have described. When you Import the file, each line of text will be placed in the cells below where you have placed the cursor. For example, the contents of the file we wish to Import is:

```
Compound      Ret.Time      Area       Amount
METHANE        3.45         87654       45.89
ETHANE         4.33         98765       98.54
PROPANE        5.66         34567       32.87
```

Note the contents of this file are the same as the one we discussed above, but the file does not contain any formatting characters like commas and quote marks. 1-2-3 has no way to distinguish where labels and numbers begin and end. With the cursor at cell A1, when this file is imported using TEXT mode, 1-2-3 treats each line as a single label and places them in the A column. Cell A1 contains the LABEL:

```
Compound      Ret.Time      Area       Amount
```

while cell A2 contains the LABEL:

```
METHANE        3.45         87654       45.89
```

and cells A3 and A4 hold the next two lines from the file. There are NO values in the B, C or D columns. We can not sum up the areas or compute anything because we have no numeric data on the spreadsheet. We have only labels.

Data Parsing to the Rescue

To rescue us from this problem, we can use the Data Parsing commands. We will parse only the body of our report in this example, though the first line of column heading can also be parsed. Data Parsing requires a few simple steps. First you must create a parsing template on one of the lines in the spreadsheet. This is easily done by placing the cursor in cell A2 on our current spreadsheet. Enter the command:

```
/Data Parse Format Create
```

A template line will be inserted in a new row just above the cursor line. This template line will contain a combination of L, V, > and # symbols indicating the position of labels and values. The contents of the row where the cursor is placed is used to create this template. If the template needs to be modified, use the command:

/Data Parse Format Modify

to add additional characters to a label or value.

Select Ranges

Now show 1-2-3 the column of cells to parse as the INPUT range and the range of cells to send the parsed data as the OUTPUT range. For our applications these ranges are:

/Data Parse Input **A2..A4**

then select Output and enter A6..D9

Then select GO. The data pared from our original labels will be placed in the output cells. Now cell A6 contains the label METHANE while B6 contains the value 3.45. Cell C6 contains the value 87654 and cell D6 has the value 45.89. Similarly, the cells in rows 7 and 8 have the labels and values for ETHANE and PROPANE. By using the Data Parse commands you can thus convert text files into usable labels and values. If you need to parse many similar files of data you can write a macro to perform the parsing.

1-2-3 Version 1A

There are no Data Parse commands in 1-2-3 Version 1A. You will have to perform the data parsing before you try to Import the file. The most direct method to perform the parsing is with a BASIC program. A BASIC program will also allow you to place the exact data you want in the order you want into the file which you will import.

A BASIC program like the one discussed in the next section is the type of program you will need to write or have written. If you are interested in learning to program, this type of project is an excellent one to get you started.

Converting Data Formats

You may have data on your disk but it is in the wrong format to import it into 1-2-3. If the original data is a straight ASCII text file or even a binary file, you can write a BASIC program to format a new file exactly the way you need the data so it can be imported into 1-2-3. If you have binary data, a program will be the only way you will be able to import data into 1-2-3. Most applications using 1-2-3 require no programming. In some cases though, there is just no convenient way to get data into 1-2-3 unless the data is formatted correctly. This is when a small BASIC program can perform the missing step very efficiently. We will develope a filter program to convert a report file into a sequential file with the right format to read into 1-2-3. This same file format will also be read into most data management programs like Paradox2, R:Base System V, and dBASE III Plus.

An Easy BASIC Program

Even if you have never written a BASIC program, a file conversion program of this type is easy to construct and implement. This type of program falls into the class of programs the programming gurus call "filters". Filter is an excellent description for the process. The input to the program is a file, the contents of the file are "filtered" and the results of the filtering are placed into a second new file Figure 3.2.

Example Program

Here is an example BASIC program which will perform the necessary format conversion. For additional background reading on BASIC programming, consult your BASIC programming Manual. In particular, read Appendix A BASIC Disk Input and Output for information on reading and writing disk files.

```
100 CLS
110 INPUT "INPUT FILENAME ?",INFILE$
120 INPUT "OUTPUT FILENAME ?",OUTFILE$
130 OPEN INFILE$ FOR INPUT AS #1
140 OPEN OUTFILE$ FOR OUTPUT AS #2
```

Prompt user for input and output file names. Then open the input file and assign it channel number 1 and open the output file and assign it channel number 2.

```
150 INPUT #1, A$
155 PRINT "Input Line is: ",A$
```

Read in the first line in the file. Then display it on the screen to show the program is working.

```
160 LABEL1$=MID$(A$,1,8)
170 LABEL2$=MID$(A$,13,8)
180 LABEL3$=MID$(A$,24,5)
190 LABEL4$=MID$(A$,35,7)
```

Figure 3.2 The structure for a typical data filtering program. Input is an ASCII text file. Each line in the text file is read and reformatted with delimiters, quote marks around labels and commas separating values.

```
INPUT

Compound          Ret.Time          Area          Amount
METHANE             3.45            87654          45.89
ETHANE              4.33            98765          98.54
PROPANE             5.66            34567          32.87

FILTER

         Prompt for INPUT filename
         Prompt for OUTPUT filename
         Read a line from INPUT file
         place Quotes around text strings
         separate numbers with Commas
         Write result line to OUTPUT file
         Close files

OUTPUT

"Compound","Ret.Time","Area","Amount"
"METHANE",3.45,87654,45.89
"ETHANE",4.33,98765,98.54
"PROPANE",5.66,34567,32.87
```

Using the text string cutting MID$ command, chop up the input text into the various labels. The MID$ command has the following structure.

```
NEW.TEXT$=MID$(TEXT$,start,characters)
```

where:

*NEW.TEXT$* is a new text string created from the original string TEXT$.
start is the number of the first character in the original text string to begin NEW.TEXT$
characters is the number of letters in the new text string.

```
200 WRITE #2,LABEL1$,LABEL2$,LABEL3$,LABEL4$
205 PRINT "Output Line is: ",LABEL1$,LABEL2$,
LABEL3$,LABEL4$
```

Write out the resulting labels to the output file. The WRITE command in BASIC will automatically place quote marks around the text strings and separate each string with commas. The output file will thus have the line:

```
"Compound","Ret.Time","Area","Amount"
```

```
210 FOR I=1 TO 3
220 INPUT #1,A$
225 PRINT "Input Line is: ",A$
230 COMP.NAME$=MID$(A$,1,10)
240 RTIME$=MID$(A$,13,7)
250 RTIME=VAL(RTIME$)
260 AREA$=MID$(A$,22,8)
270 AREA=VAL(AREA$)
280 AMOUNT$=MID$(A$,33,8)
290 AMOUNT=VAL(AMOUNT$)
300 WRITE #2,COMP.NAME$,RTIME,AREA,AMOUNT
305 PRINT "Output Line is:
",COMP.NAME$,RTIME,AREA,AMOUNT
310 NEXT
320 CLOSE
330 END
```

Within a FOR .... NEXT loop the next three lines in the INPUT file are read in, and chopped up with the MID$ command. The numeric values are then converted from text string representations into numbers with the VAL function. The VAL function has the following structure:

```
number = VAL (string$)
```

where the function places the numeric value of the string$ in the variable number.

Each line of the INPUT files is then converted into a string and three numeric values. Using the WRITE command, the string is automatically delimited with quote marks and the numbers are separated by commas. The first data line in the new file will read:

```
"METHANE",3.45,87654,45.89
```

Our program ends by closing the data files and ending.

Using the Program

This particular program of course will not perform the file conversion you will need for your particular application. The program does spotlight the information you will need to write your own program like the structure of the INPUT file. You will have to get right in there and count the number of characters there are in each line of the INPUT files and decide where to make the cuts. In this example program all of the data from the INPUT file was sent to the OUTPUT file. In your program you may want to send just specific pieces of data and possibly change the order of the data.

```
         A         B         C         D         E         F         G         H

1  Read lines from a file
2
3  \r        {GETLABEL "File to read (include drive and ext): ",filename}
4            {GETNUMBER "Enter number of lines to read: ",num_lines}
5            {OPEN filename,R}
6            /RNChere~E10~
7            {FOR counter,1,num_lines,1,readwhat}
8            {CLOSE}
9            /XG\R~
10
11 readwhat  {READLN A5}
12           {GOTO}here~/MA5~~
13           /RNDhere~{DOWN}~
14           /RNChere~~
15           {RETURN}
16
17 counter            6
18 filename  b:test.prn
19 num_lines          5
20
```

Figure 3.3 Lotus macro for reading ASCII files directly from disk.

Importing Data using a {READLN} Macro

The macro {READLN} command can read a line of text from an ASCII file. This can be a very useful command if you do not wish to load an entire file in order to obtain certain pieces of data. The macro shown in Figure 3.3 reads a specified number of lines from the selected file. Each line is then saved starting at line E10. To use this macro type in the labels shown in the A column and the macro commands in the B column. The data in cells B17, B18 and B19 does not have to be entered since it will be entered by the responses you give when running the macro.

After entering the labels and commands, assign the range names with the command:

/Range Name Labels Right **A3..A19**

The macro is now ready to run by pressing Alt-r. Answer the prompts and read in an ASCII file.

Macros like the one shown can be modified to read a single line of data, remove a specific single value or label from the line and then continue

Figure 3.4 By editing
the worksheet only
specific parts of the disk
file will be saved in the
worksheet. This saves
worksheet space and data
loading time.

```
Add to cell A6: @MID(A5,Ø,2Ø)

This keeps only the first 20 characters.

Edit cell B12: {GOTO}here~/RVA6~~

This uses the /Range Value command to move the data.
```

to the next line. You can use string commands to capture the characters
you wish to keep and then copy only the parts of the data you wish to
keep.

By modifying the "readwhat" part of the macro as shown in Figure 3.4,
only the first label and value in each line is saved.

Importing Data using RAM Resident Programs

RAM resident programs like Borland's SideKick and Lotus Metro can be
used to transfer data from other programs or files into 1-2-3. Either of
two methods can be used depending on your situation. Both methods use
the RAM resident program's capability to "play-back" selected text as
if it was being typed.

Cutting and Pasting a Report

If you need to create reports with the results from a number of different
files SideKick is very helpful. You can run your wordprocessor and start
creating the report. When you get to the location in your report where
you need information from another file, press Ctrl-Alt and view the
actual contents of the results file you need on the screen in SideKick's
notepad as shown in Figure 3.5.

Select the Text You Want

Now you can select the part of the results file you need by using the Text
Block selection commands. Move the cursor to the start of the Text Block
and press Ctrl-K then B, to identify the beginning of the block. Then
move the cursor to the end of the block of text you need and press Ctrl-K
and K.

Transfering the Text

You transfer this block of text into your wordprocessor or other programs
by using the cut and paste commands. You do this by assigning a special
key combination, a magic key, to the selected block of text. After
exiting SideKick, each time you press the special key combination, the
text you selected will be played back into the program you are then
running.

```
          A          B          C          D          E          F          G          H
1
2
3
4
5
6
      B:\100.RPT                      Line 22    Col 1    Insert      Indent

  NAME            LB/GAL         RT      AREA BC     RF        RRT

  COMPONENT1          0.132      0.33   132364 01    1000.       0.5
  COMPONENT2      132358.        0.66   132358 01       0.001   1.
  COMPONENT3      123.           0.99   132348 01       1.076   1.5
  COMPONENT4      202.           1.33   132382 01       0.655   2.015
  COMPONENT5      203.           1.33   132382 01       0.655   2.015
  COMPONENT6      204.           1.33   132382 01       0.655   2.015
  COMPONENT7      205.           1.33   132382 01       0.655   2.015

  TOTALS          132683.132            529452
```

F1-help F2-save F3-new file F4-import data F9-expand F10-contract Esc-exit

Figure 3.5 SideKick's notepad displays the text file.

Selecting the Magic Key

Press Ctrl-K then E, you are then prompted for the special key combination. Let's use the key combination Alt-A. Then you are asked if you want to have the text played back in Block mode or Line mode. Block mode plays back the entire text block you have selected while Line mode plays back one line at a time and you must press the Enter key. Line mode is useful if you need to place the text in various locations in your document.

Use Multiple Magic Keys to Transport Data

The cut and paste commands are also useful in transporting data from your reports into other programs such as 1-2-3 and data managers. When you press the magic key, it is just as if you are typing in the characters. You can save different portions of a results file with different magic keys. Then you can play back those results in the program where you need them. This provides a simple yet powerful way to save time consuming re-keying of data.

Using the techniques described in this chapter you should be able to get any data which is on disk into a 1-2-3 worksheet.

Figure 3.6 Simple
report to be exported from
Lotus 1-2-3.

```
Pk# Compound Ret Time      Area    Amount
10  Methane      2.34     12345    52.630
11  Ethane       2.67     34567    99.936
12  Propane      3.68     78904   336.076
13  Butane       4.98      5674    23.954
14  Pentane      6.89     12356    50.303
15  Hexane       9.78    896654  3334.030
16  Heptane     10.76      9223    39.319
17  Octane      13.45     14568    59.299
```

Exporting Your Results

Sharing Lotus 1-2-3 data with other programs can be easily performed by using one or more of the following exporting data techniques. The exported data can not only be read by other programs but also your own BASIC programs.

For the examples a simple chromatography data report will be exported from Lotus 1-2-3. The report, shown in Figure 3.6, has five columns of data, peak number, compound name, retention time, compound area and amount. The compound amount is computed from the compound areas by using an external standard.

Here are four different ways you can export your Lotus 1-2-3 data to other programs.

Fixed Length Field ASCII Files

A common file format is to place data in ASCII files with fixed length fields. The example data in this file format looks like the data shown in Figure 3.7. These types of files can be easily read into data management programs for entry into a database. These files can also be read by BASIC programs like the one described later in this section. A fixed length field file places its data in individual lines with a specific number of characters or character spaces reserved for each piece of data. In our chromatography report we have four columns or fields of data. For this example we want to have a 20 character wide column for the compound names, 10 character wide column for the retention times, 15 character wide column for the areas and a 12 character wide column for the amount.

These types of files can be generated from your data by using the /Print File command. The width of each worksheet column will determine the number of characters reserved for each field. So before we use the Print

```
10   Methane    2.34    12345    52.630
11   Ethane     2.67    34567    99.936
12   Propane    3.68    78904   336.076
13   Butane     4.98     5674    23.954
14   Pentane    6.89    12356    50.303
15   Hexane     9.78   896654  3334.030
16   Heptane   10.76     9223    39.319
17   Octane    13.45    14568    59.299
```

commands we must use the /Worksheet Column Set-Width command to set the column widths we desire. Then use the /Range Format command to select the number of decimal places you wish to transfer. Remember the data is being transformed into ASCII characters, so the values you see on the worksheet screen will be the ones you transfer. Depending on how you format your columns, the data you transfer will be rounded to the number of decimal places you select.

When using the Print File command for this type of data transfer, be sure to actuate two commands. First, make sure the Options Margins Left and Options Margins Right span enough characters to include the full width of your data. A maximum of 240 characters per line is available. If you have more than 240 characters per line, you will have to divide up your data and place it in two or more different files. Actuate the the Options Other Unformatted mode. This command turns off all page formatting so that only the worksheet data is stored, not any header or footer lines.

Comma Delimited Files

In a comma delimited file each piece of data is separated from other data with a comma. A comma delimited file for our example data would look like the data shown in Figure 3.8. Numbers are simply separated by commas while text also has quotation marks around each entry.

These files are most easily created by using a Lotus macro using the {WRITE} command. Such a macro is shown in Figure 3.9. To use this macro, enter the macro as shown. Enter the functions as shown. Place the cursor at the first line to export and press Alt-E. The macro will output each line until it reaches a blank cell.

RAM Resident Program

RAM resident programs like SideKick and Lotus Metro can also be used to transfer data from Lotus 1-2-3 to other programs. Most of these programs have the ability to capture the text displayed on your computer

Figure 3.8 Comma delimited file. Each piece of data is separated by commas. Text is surrounded with double quote marks.

```
10,"Methane",2.34,12345,52.630
11,"Ethane",2.67,34567,99.936
12,"Propane",3.68,78904,336.076
13,"Butane",4.98,5674,23.954
14,"Pentane",6.89,12356,50.303
15,"Hexane",9.78,896654,3334.030
16,"Heptane",10.76,9223,39.319
17,"Octane",13.45,14568,59.299
```

screen. Usually the text can be stored in a text file which can be later printed or combined with other text using a wordprocessor. Here is how text on your 1-2-3 screen can be captured using SideKick.

First load SideKick into your computer's memory, load Lotus 1-2-3 and load in our example with these commands.

```
SK
123
/File Retrieve EXDATA
```

Now press Ctrl-Alt or the "hot-key" combination you have selected for SideKick. Select the notepad and then open a new file by pressing [F3] and entering the name TRANSFER. Select the import option by pressing [F4]. The SideKick window will disappear leaving the normal 1-2-3 worksheet on the screen. A blinking single line cursor will be in the upper left corner of the screen, use the arrow keys to position the cursor at the beginning of the text data in cell A4. Press the keys Ctrl-KB, this marks the upper left corner of the text you wish to transfer. You can use the arrow keys to select the text you wish to transfer. The text to transfer will be highlighted on the screen as you make the selection. Highlight the area on the screen containing the sample data so that it covers to the lower right corner of cell D9. The press the keys Ctrl-KK, this indicates the lower right corner of the text you wish to transfer.

After pressing Ctrl-KK the SideKick window will reappear with a blinking cursor showing in your notepad display. Move the cursor to where you wish to have the screen text and press Ctrl-KC. The screen text will then be transfered into the notepad. Save the text by pressing [F2]. This text can then be edited or combined with other text using your normal wordprocessor.

```
        A       B       C       D       E       F       G       A3: 'S80976
1   Write a comma delimited file to disk                            B3: 98.76
2                                                                   C3: 'Gold
3   S80976         98.76 Gold              7                        D3: 7
4   "S80976",98.76,   "Gold",  7.0                                  A4: @CHAR(34)&@TRIM(A3)&@CHAR(34)&","
5   "S80976",98.76,"Gold",7.0                                       B4: @TRIM(@STRING(B3,2))&","
6                                                                   C4: @CHAR(34)&@TRIM(C3)&@CHAR(34)&","
7   \w      {GETNUMBER "Number of lines to Write: ",numdata}        D4: @TRIM(@STRING(D3,1))
8           {GETLABEL "File (include disk & ext): ",filename}       A5: +A4&B4&C4&D4
9           {GOTO}a27~
10          {OPEN filename,W}
11          {FOR counter,1,numdata,1,wrtline}
12          {CLOSE}
13
14  wrtline /C~A3~{RIGHT}
15          /C~B3~{RIGHT}
16          /C~C3~{RIGHT}
17          /C~D3~{DOWN}{LEFT 3}
18          {WRITELN A5}
19          {RETURN}
20
21  numdata         5
22  counter         6
23  filename b:text.prn
24
25          Instrument
26  Sample  Reading Color           Grade
27  S90723     45.67 Tan               1
28  S95124     34.56 Yellow            4
29  S95118     23.45 Red               3
30  S93467     45.32 Blue              2
31  S80976     98.76 Gold              7
32
33
34  "S90723",45.67,"Tan",1.0
35  "S95124",34.56,"Yellow",4.0
36  "S95118",23.45,"Red",3.0
37  "S93467",45.32,"Blue",2.0
38  "S80976",98.76,"Gold",7.0
```

Figure 3.9 Lotus macro using the {WRITE} macro command to generate comma delimited files.

Reading the Lotus File Formats

Lotus worksheet and graphics files can also be read directly if you have programs which can read the Lotus file formats. The file formats have been published by Addison-Wesley in the book "Lotus File Formats for 1-2-3, Symphony & Jazz".

4

Presenting Your Results

Once your data is in Lotus 1-2-3, there are many ways you can print and plot your results. Worksheet reports and tables can be printed directly to your printer or to a file. Lotus graphs are plotted using the PGRAPH program. For more interesting results you can transfer your data to other programs to enhance the graphics, combine results with other text or other graphs.

Laser Printing in the Lab

The newest technology to become cost effective for permanent graphic copies is laser printing. This method of graphical documentation holds out the most promise for laboratory applications because it can combine high-resolution "near typeset quality" text characters with high-resolution graphs. Laser printer prices are rapidly moving down while software to easily combine both text and graphics on a page is rapidly improving. PC-generated laser-printed reports are already being used routinely in laboratories. Laboratories where forms are extensively used can take advantage of a laser printer's ability to print both the form with the data in one rapid step.

Most laser-printers like the Hewlett-Packard LaserJet and the Apple LaserWriter have resolutions of 300 dots per inch. Common dot-matrix printers have resolutions of about 75 dots per inch while true typesetting machines used in publishing magazines and books like the Allied Linotype Linotronic 100 and 300 imagesetters provide 1000-2400 dots per inch resolution (Figure 4.1)

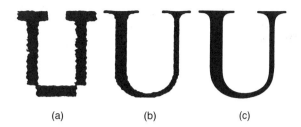

Figure 4.1 Comparison of available printing technology. A good dot matrix printer (a) has resolution of about 75 dots per inch, a laser printer (b) 300 dots per inch and a typesetting machine (c) provides from 1000 to 2400 dots per inch.

(a) (b) (c)

Enhancing 1-2-3 Graphs with Freelance Plus

Lotus Freelance Plus can not only create on-screen charts and graphs but it can also edit them. If you created a graph or chart using Lotus 1-2-3 it can be edited and enhanced with Freelance. With Freelance you create and edit the graphical images directly on the screen. This makes Freelance extremely easy to use while the quality of the graphics will fill any presentation graphics needs.

Freelance is an object-oriented, rather than a pixel-oriented program. This means images are stored as objects formed by lines and arcs (sometimes called vectors). Each line in an object is defined to be a specific length going in a specific direction.

Other drawing programs, such as Micrographx Windows DRAW!, also uses object-oriented image storage and work with Microsoft Windows. Object oriented symbol representation and storage allows you to easily make any image larger or smaller without losing any resolution or detail. The final output is constrained only by the resolution of the output device. Pixel-based graphics programs can not do this since they are constrained by the resolution of the computer screen. When images are enlarged using a pixel-based drawing program, the program can only illuminate additional pixels.

Most other drawing programs are pixel-based meaning they control which pixels (dots or pels) on your graphics monitor are to be on, off or display a specified color. Pixel-based programs are constrained to the resolution of the computer screen. This resolution is fine if you are preparing a computer slide-show, but it is not adequate if you want high quality presentation graphics produced using a plotter, laser printer or film recorders.

Using Freelance

Freelance is very easy to use. The program has three screen areas, the main working area, the command panel at the top of the screen and a

Figure 4.2 Freelance display with graphics work area, command and control panels.

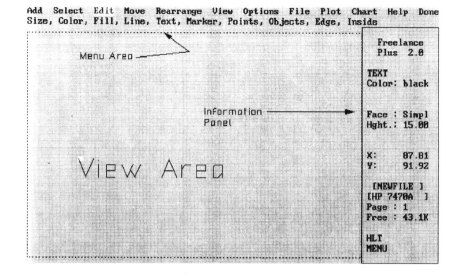

status panel on the right side (Figure 4.2). The program uses a menu selection structure similar to Lotus 1-2-3. You can make a command selection by moving the cursor over the command you wish to use and press the enter key or you can enter the first letter of the command. All of the graphic composition is performed on the screen so you always see exactly what you are working on. Text or drawn objects can be edited, moved, replicated, rearranged, flipped and deleted with just a few key strokes or movements of a mouse. To create precise figures a grid can be displayed with a user defined grid size. The status panel also shows the X and Y position of the cursor. The X Y coordinates can also be set with a RULER command to represent any linear units so you can make the units angstroms for example if you are drawing molecules to scale or miles if you are drawing maps.

Symbol Storage and Libraries

A major feature which makes Freelance a unique software product is the ability for you to create and use libraries of symbols. Freelance comes with two disks full of symbols including maps of the United States, all the state maps, geometric shapes, flow charting aids and symbols, common office and industrial objects like computers, office furniture, signs, trains and airplanes (Figure 4.3). These symbols can be easily transferred from a library file to your working area.

Using Library Symbols

Library symbols are stored in disk files. To use a symbol, you first switch to the second viewing screen with the VIEW SWITCH command. Then you read in the symbol file with the symbol you wish. Then you switch back to the primary working screen with the VIEW

Figure 4.3 Examples of symbols which come as part of Freelance.

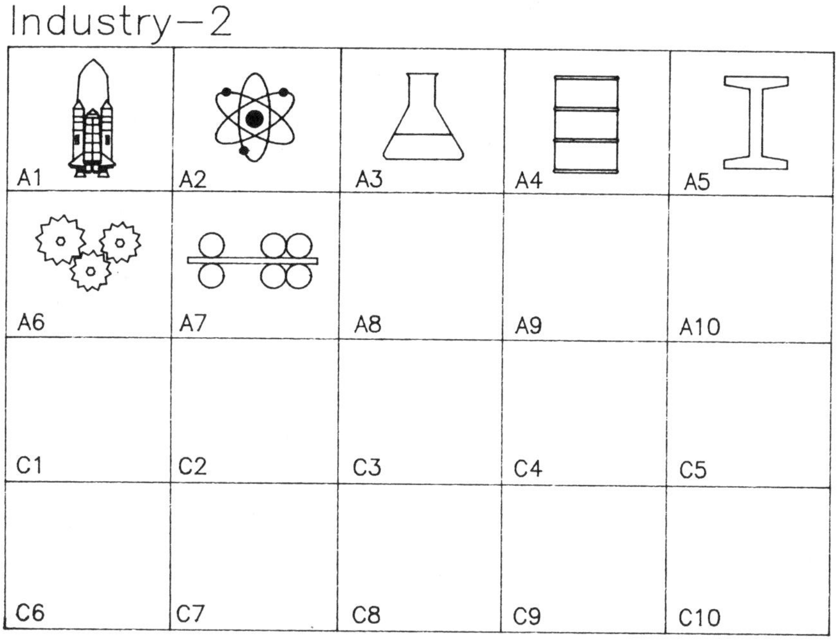

Industry—2

A1	A2	A3	A4	A5
A6	A7	A8	A9	A10
C1	C2	C3	C4	C5
C6	C7	C8	C9	C10

SWITCH command. The symbol files are divided into twenty different storage areas (Figure 4.4). The contents of a storage area can be transfered into the primary screen by using a combined keystroke. You use the Ctrl key and one of the ten function keys for the first ten areas or the Alt key and one of the ten function keys for the next ten areas.

Additional symbol libraries are being produced by third party vendors to work with Freelance. For example, maps including counties or major highways are available. Even libraries of chemical structures are available as shown in Figure 4.4!

With Freelance you can create any type of chart, graph, slide or overhead transparencies for presentations or publications. Freelance is an excellent presentation graphics product.

Combining Text

Here are four different ways you can combine text with your Lotus 1-2-3 results. The methods will proceed from the simplest and least costly to the most complex and expensive.

Using a Stand-alone Wordprocessor

Data you have in your Lotus worksheets can be combined with text right in your own wordprocessor. Using any of the techniques described in the "Exporting Lotus 1-2-3 Data" section of chapter 4, you can store your data in a text file. The text file can then be read into your wordprocessor, edited, combined with other text and printed.

Figure 4.4 A page of structures from the Chemical Structures for Freelance. Twenty different structures or symbols can be placed in one disk file.

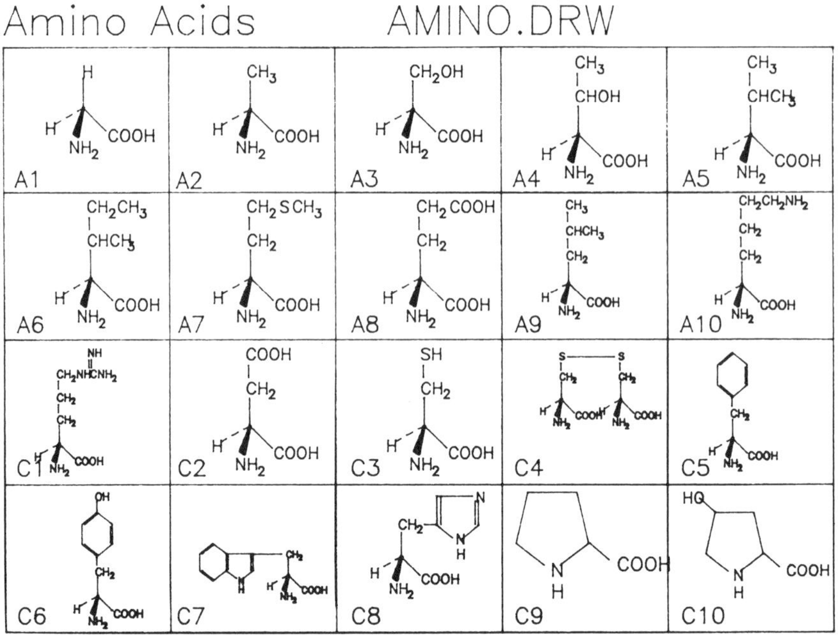

To combine a graph with your text, you must leave spaces in the final copy and paste in the graphs after you have printed or plotted them.

Lotus Add-in Wordprocessors

A number of third-party Lotus add-in programs provide wordprocessing capabilities right within your 1-2-3 worksheet. All of these add-in programs give you the ability to integrate text with your 1-2-3 data and most provide "mail-merge" capabilities with Lotus 1-2-3 databases. Some can even print graphs with text.

Three of the more popular add-in wordprocessors are:

Write-in
Blossom Software Corp
One Kendall Square, Suite 2200
Cambridge, MA 02142
800 852-8017 in MA 617 666-2144
Price: $99.95

InWord
Funk Software
222 Third Street
Cambridge, MA 02142
617 497-6339
Price: $99.95

4WORD
Turner-Hall Publishing
10201 Torre Avenue
Cupertino, CA 95014
(408) 253-9600
Price: $99.95

Lotus Manuscript

Lotus Manuscript is a word and document processor designed especially for complex documents such as reports and technical proposals. It uses the familiar 1-2-3 user interface to select commands. Manuscript has a built in outliner and spelling checker. Manuscript can read a Lotus 1-2-3 worksheet directly as well as graph (.PIC) files. Manuscript can then mix text, tables, charts, diagrams and equations all on the same page. Equations can be sized and can include greek symbols. It can handle documents up to 800 pages long and can provide a table of contents, tables of figures, tables of tables, indexing, cross referencing and floating footnotes.

A number of printers including laser printers are supported by the program. Before a page is printed, it can be previewed for proper positioning with the document preview feature as shown in Figure 4.5.

Aldus PageMaker

Aldus Pagemaker is a page composition program. The program provides all the tools a graphics designer needs to combine text and graphics to create an eye appealing presentation of the information. Text is provided to the program in text files created on a wordprocessor. The text output from all of the major wordprocessing programs is supported. ASCII text files are also supported. Graphics can be obtained from many different sources including Lotus graph (.PIC) files. Graphics captured from scanners and also other graphics programs like PC-Paintbrush, and Windows Paint are also supported.

Lotus worksheets can be used in Aldus Pagemaker by first exporting the worksheet data into an ASCII text file. The ASCII text file can then be placed in its proper location. Numerous font styles and sizes can be selected for each piece of text. Lotus graph files can be read directly and placed on a page.

PageMaker uses a "What-You-See-Is-What-You-Get" user interface. The current page composition is always shown on the screen. Portions of the page can be viewed in actual size and even 200% size. Overviews can be obtained by viewing a page at 50% or 75% actual size. Facing pages can also be displayed side-by-side. Complete pages are printed on a laser printer or for more resolution on a Linotype high resolution printing system.

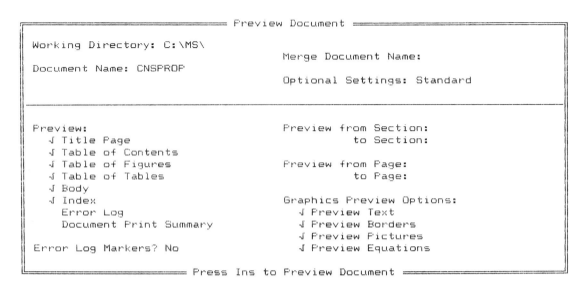

Figure 4.5 Example of a Manuscript preview screen.

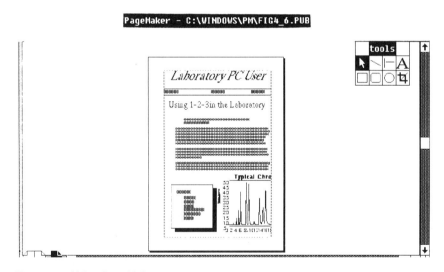

Figure 4.6 Aldus PageMaker screen.

Products mentioned in this chapter:

Freelance Plus, Lotus Development Corp. 55 Cambridge Parkway, Cambridge, MA 02142 list price $495.

Windows DRAW!, Micrographx Inc. 1820 N. Greenville Ave. Richardson, TX 75081 (214) 234-1769 list price: $199

Chemical Structures for Freelance, BREGO Research 5989 Vista Loop San Jose, CA 95124 (408) 723-0947 list price: $99

Aldus PageMaker, Aldus Corp. 411 First Avenue South, Seattle, WA 98104 (206) 622-5500, list price: $695

Lotus Manuscript, Lotus Development Corp. 55 Cambridge Parkway, Cambridge, MA 02142 list price $495.

PART TWO: Laboratory Applications

These chapters contain 10 applications of Lotus 1-2-3. The applications start with reporting results, then methods for generating calibration curves and fitting experimental data are described. Data analysis techniques using graphics are then highlighted followed by applications for summarizing and managing data.

Chapters 5 through 8 are introductory applications of Lotus 1-2-3. You should be able to perform each of them in 15 to 20 minutes at the most from start to finish. These applications will introduce you to Lotus 1-2-3 and provides examples for using the Lotus spreadsheet, graphics, and data management commands.

The first application, chapter 9, demonstrates how data can be analyzed and reported using 1-2-3. A chromatography example is used. The reporting worksheets generated in that application can be used to report any multi-component analysis.

The next three applications, chapters 10, 11, and 12, deal with fitting curves or lines to data using the Data Regression command. The original data and the regression curves or lines are then plotted. The potential error in a linear fit is then graphically displayed by plotting the confidence interval for a calibration line.

Chapters 13 and 14 are applications of 1-2-3 data analysis. In chapter 13 a 3D plotting add-in is used to create 3D graphs of worksheet data. In chapter 14, semi-log plots are generated to more easily analyze data.

Chapter 15 shows how macros can be used to load the contents of many individual reports. This data can then be summarized techniques described in chapters 16, 17, 18, and 19.

The final four chapters, 16, 17, 18, and 19 are applications which use Lotus Data Management and graphics commands to plot, summarize, and track analysis results.

5

Testing Worksheet Precision and Range

This application will teach you how to use some basic Lotus 1-2-3 commands and also verify the precision and range of the numbers you can use in the program.

One of the first things you should always do when you start working with a new program is to find out or verify the precision and the dynamic range of numbers which can be represented in the program. This is particularly important when working on scientific applications since values you work with can be very large or very small.

Decimal Places

The precision of numbers is usually measured by the number of decimal places which can be stored and used in the program. Lotus 1-2-3 can store numbers with up to 15 decimal places. The control panel, however, displays a maximum of 9 decimal places. For example, if you enter a number with 12 decimal places, 1-2-3 will perform calculations using all 12 decimal places but you will see only 9 decimal places on the control panel and up to 10 decimal places for values on the worksheet. If more decimal places must be represented, the number is displayed in scientific notation. For large numbers, up to 15 decimal places will be displayed before they are displayed in scientific notation.

Dynamic Range

The range of numbers which can be represented is also important. Lotus 1-2-3 can accept numbers in the range from 10^{-99} to 10^{99}. 1-2-3 can store numbers as a result of calculations ranging from 10^{-308} to 10^{308}.

Both the precision and dynamic range of numbers you can use in 1-2-3 is very good and more than acceptable for most scientific and engineering applications. The reason for this good precision and large range is that these capabilities are needed not only for scientific applications but also for precise currency exchanges!

Let's test these claims while also learning a few command applications. We will create a worksheet which will test the precision and range of numbers in Lotus 1-2-3.

Testing Small Values

Load Lotus 1-2-3 and make sure you are starting with a clean worksheet by entering the command:

```
/Worksheet Erase Yes
```

We will test the precision and dynamic range by setting up a worksheet which has very large and small values. Precision will be tested by viewing the number of decimal places displayed. Dynamic range will be tested by seeing the range of values which can be displayed. The large range of values will be created by using the copy command to copy a formula to a column of cells.

Enter these values and formulas in these cells:

```
B1: +1/3   B2: +b1*.1
```

Now copy the formula in cell b2 to the range b3 to b100 with the command:

```
/Copy b2..b2  [Enter]  b3..b100  [Enter]
```

This column of values should range from .333333333 to 3.33 E+100. Notice the precision of the values is shown by the number of decimal places displayed. To inspect the range of values, move the cursor to cell B1 then press {END}{DOWN}. The cursor should be in cell B100. Notice the lowest value displayed is 3.3E-99. The next cell is filled with *****. This is the indication 1-2-3 uses to show a value or display format is too large for the current column width or the value is out of the dynamic range of the program. Change the column width to 15 spaces with the command:

```
/Worksheet Column Set-Width 15 [Enter]
```

Notice that cell B100 is still filled with ****, indicating the value in the cell is too small to represent on the worksheet.

Testing Large Values

We will use a similar test to display large values. Enter these values, formulas and commands:

```
C1:100/3
C2:C1*10
/Copy from C2..C2 [Enter] C3..C100 [Enter]
/Worksheet Column Set-Width 15 [Enter]
```

Note the largest value displayed before scientific notation is used has 15 numbers. Now

```
Move to: C100
```

Notice the large numbers end with 3.33 E+99. The next cell is filled with ****, indicating it can not be represented.

The final worksheet will look like Figure 5.1.

Figure 5.1 Worksheet
testing the resolution and
dynamic range of numbers
used in Lotus 1-2-3.
Pass/Fail Report

	A	B	C
1		0.3333333333	33.333333333
2		0.0333333333	333.33333333
3		0.0033333333	3333.3333333
4		0.0003333333	33333.333333
5		0.0000333333	333333.33333
6		0.0000033333	3333333.3333
7		0.0000003333	33333333.333
8		0.0000000333	333333333.33
9		0.0000000033	3333333333.3
10		0.0000000003	33333333333
11		3.3333333E-11	333333333333
12		3.3333333E-12	3333333333333
13		3.3333333E-13	33333333333333
14		3.3333333E-14	3.3333333E+14
15		3.3333333E-15	3.3333333E+15
16		3.3333333E-16	3.3333333E+16
17		3.3333333E-17	3.3333333E+17
18		3.3333333E-18	3.3333333E+18
19		3.3333333E-19	3.3333333E+19
20		3.3333333E-20	3.3333333E+20
90		3.3333333E-90	3.3333333E+90
91		3.3333333E-91	3.3333333E+91
92		3.3333333E-92	3.3333333E+92
93		3.3333333E-93	3.3333333E+93
94		3.3333333E-94	3.3333333E+94
95		3.3333333E-95	3.3333333E+95
96		3.3333333E-96	3.3333333E+96
97		3.3333333E-97	3.3333333E+97
98		3.3333333E-98	3.3333333E+98
99		3.3333333E-99	3.3333333E+99
100		**************************

6

Pass/Fail Report

To insure crisp crackers, the moisture content of sample crackers is measured after they have been baked and cooled. Before the crackers are packaged a representative sample of each batch is tested. If the moisture content is greater than 10 percent, the entire batch is heated for an additional 15 minutes to drive off any additional moisture.

The analysis is performed by first weighing the cracker sample, heating it in a hot oven for 2 minutes and then after cooling the sample is weighed again. The difference in weight is assumed to be water. To calculate the percent water, the following calculation is made:

(Ending Weight - Starting Weight) * 100 / Ending Weight

Samples which have greater than 10 percent moisture are give a fail status while those with less than 10 percent moisture are passed.

This is a very simple calculation, but if you must perform 30 of these tests every hour and report the results quickly and accurately, the data analysis and reporting can become tedious. Let's set up a 1-2-3 worksheet to perform this task.

Start with a clean worksheet by entering the command:

    /Worksheet Erase Yes

Enter these labels for the report:

```
A1: Pass/Fail Moistrue Content Worksheet
B2: Start Wt.
C2: End Wt.
D2: Moisture
E2: Batch No.
F2: PASS/FAIL
```

To keep the labels from running into each other, change the default column width for the worksheet with the command:

```
/Worksheet Global Column-Width Set-Width 10
[Enter]
```

Normally the sample weights and batch numbers would be entered by hand. For this example, we will use the Data Fill command to fill in a set of starting and ending weight values.

```
/Data Fill B3..B20 [Enter] 2.45 [Enter] .5
[Enter] 8192 [Enter]
/Data Fill C3..C20 [Enter] 2.4 [Enter] .3
[Enter] 8192 [Enter]
/Data Fill D3..D20 [Enter] 100 [Enter] 1
[Enter] 8192 [Enter]
```

Now enter the formula for the analysis result and copy it down the column:

```
D3:+(B3-C3)/C3
/Copy D3..D3 [Enter] D4..D20 [Enter]
```

Display this data in percentage format by using the Range Format command:

```
/Range Format Pecentage D3..D20 [Enter]
```

The PASS/FAIL Designation

The PASS/FAIL column can be filled by using the @IF function. The function allows you to perform a test and depending on if the results are true or false, a different label or value will occupy the cell. The form of this command is:

```
@IF (condition, true, false)
```

If the condition is true, the label or value in the true position will occupy the cell. If the condition is false, the label or value in the false position will occupy the cell.

Lotus in the Lab

For this application the following formula can be used and copied:

```
F3:+@IF(+D3<.10,"PASS","***** FAIL *****")
/Copy F3..F3  [Enter]  F4..F20  [Enter]
```

The resulting report will look like Figure 6.1.

Note the Batch Number column is placed right next to the PASS/FAIL column. This makes reading the report much easier. Those batches which do fail, are marked with many ****'s so they will stand out on the report.

Keeping Records

The results of this test can be printed on the printer and sent to the cracker packaging area. Then the results can be stored on disk. If at a future time the results must be reviewed, they will be available. To start a new analysis, simply erase the starting and ending weights with the command:

```
/Range Erase B2..C20  [Enter]
```

Now you can enter the new data and store the results under a different file name.

	A	B	C	D	E	F	G
1	Pass/Fail	Mositure	Content	Worksheet			
2		Start Wt.	End Wt.	Moisture	Batch No.	PASS/FAIL	
3		2.45	2.4	2.08%	100	PASS	
4		2.5	2.43	2.88%	101	PASS	
5		2.55	2.46	3.66%	102	PASS	
6		2.6	2.49	4.42%	103	PASS	
7		2.65	2.52	5.16%	104	PASS	
8		2.7	2.55	5.88%	105	PASS	
9		2.75	2.58	6.59%	106	PASS	
10		2.8	2.61	7.28%	107	PASS	
11		2.85	2.64	7.95%	108	PASS	
12		2.9	2.67	8.61%	109	PASS	
13		2.95	2.7	9.26%	110	PASS	
14		3	2.73	9.89%	111	PASS	
15		3.05	2.76	10.51%	112	***** FAIL *****	
16		3.1	2.79	11.11%	113	***** FAIL *****	
17		3.15	2.82	11.70%	114	***** FAIL *****	
18		3.2	2.85	12.28%	115	***** FAIL *****	
19		3.25	2.88	12.85%	116	***** FAIL *****	
20		3.3	2.91	13.40%	117	***** FAIL *****	

Figure 6.1 Pass/Fail report generated using Lotus 1-2-3.

7

Reporting Means, Standard Deviation and High-Low Ranges

To study the variability of how long it takes an upset stomach powder to dissolve in water, ten different samples of an upset stomach powder were dissolved. The amount of time, in seconds, required to dissolve each sample was as follows:

50.7	69.8
54.9	53.4
54.3	66.1
44.8	48.1
42.2	35.5

We need to report the mean, the standard deviation, the high and low observed values for this test. Start with a clean worksheet by entering:

    /Worksheet Erase Yes

Enter the title:

    A1: 'Time to Dissolve an Upset Stomach Power

Enter the observed values in column B starting with cell B2.

B9: 50.7	B15: 69.8
B10: 54.9	B16: 53.4
B11: 54.3	B17: 66.1
B13: 44.8	B18:48.1
B14: 42.2	B19:35.5

Enter these labels in column A:

```
A3:  'Average:
A4:  '    High:
A5:  '     Low:
A6:  'STD (n):
A7:  '   (n-1):
```

First we will give the range of cells containing the data a name. Naming a range of cells is a very key part of constructing good templates or worksheets. By creating named ranges of common data, you can use the range name in formulas and functions rather than the actual range of cells. If you give descriptive names to ranges, they will help document your calculations and make them easier to read and understand. More importantly, if you need to change the number of data points you want to consider, you simply redefine the cells in the named range. You do not have to redefine the range in each equation and formula. Name the range STOMACH_TIME with the command:

```
/Range Name Create  STOMACH_TIME  [Enter]
B9..B18  [Enter]
```

Now enter these functions in the B column:

```
B3:+@AVG(STOMACH_TIME)
B4:+@MAX(STOMACH_TIME)
B5:+@MIN(STOMACH_TIME)
B6:+@STD(STOMACH_TIME)
B7:+@SQRT(@COUNT(STOMACH_TIME)/
      (@COUNT(STOMACH_TIME)-1))*@STD(STOMACH_TIME)
```

The results for each of the computed values will be displayed in their respective cells. Note the @STD function computes computes the standard deviation of population data which divides by n rather than (n-1). To perform the n-1 method the equation in cell B7 must be used. The resulting worksheet will look like Figure 7.1.

Performing What-if

If any of the data values are changed, the functions and equations will instantly change. For example, if the final value is changed from 35.5 to 30.5, each of the computer values will change except for the high value.

If you would like to add an additional value to the data set, first redefine the named range to include cell B12 with the command:

Figure 7.1 Worksheet displaying the data with the computed average, high, low and standard deviations. Note the statistical data is located above the data so additional data can be easily added to the data set.

```
              A         B        C        D        E
   1    Dissolving Time for Upset Stomach Powder
   2
   3    Average:      51.98
   4       High:      69.8
   5        Low:      35.5
   6    STD (n): 9.840406
   7       (n-1): 10.37269
   8
   9                  50.7
  10                  54.9
  11                  54.3
  12                  44.8
  13                  42.2
  14                  69.8
  15                  53.4
  16                  66.1
  17                  48.1
  18                  35.5
  19
```

/Range Name Create STOMACH_TIME [Enter]
B2..B12 [Enter]

Then enter the new value in cell B12:

B12: 37.6

Notice all of the computed values are immediately updated.

This application shows how you can perform statistical calculations on any set of data you have entered on a worksheet.

8

Sorting and Plotting Experimental Data

The sorting and plotting capabilities of Lotus 1-2-3 will be used to analyze some experimental data.

Five different rats were timed as they were run through four different mazes. The times for each rat running each maze are shown below.

To analyze this data, start with a clean worksheet by entering the command:

/Worksheet Erase Yes

Then enter the titles and column headings:

```
A1:  'Rat Maze Times
B2:  'Rat Name
C2:  'Maze 1
D2:  'Maze 2
E2:  'Maze 3
F2:  'Maze 4
```

Then enter the rat names and their times:

```
B3:  'Trigger    C3: 78    D3: 103    E3: 134    F3: 146
B4:  'Slew       C4: 86    D4: 99     E4: 145    F4: 135
B5:  'Citation   C5: 88    D5: 112    E5: 155    F5: 154
B6:  'Bold       C6: 81    D6: 102    E6: 168    F6: 155
B7:  'Dancer     C7: 85    D7: 116    E7: 157    F7: 156
```

Then compute the average time for each rat through all of the mazes:

```
G3: +@AVG(C3..F3)
/Copy G3..G3  [Enter]  G4..G7
```

The worksheet should look like Figure 8.1.

Sorting the Data

Now, the data can be sorted alphabetically or by number. Let's sort alphabetically first. Enter the following commands:

```
/Data Sort Data-Range B3..G7  [Enter]
Primary-Key B3 [Enter] A Go
```

The data will be sorted into alphabetical order using the B column. The result of the sort is shown in Figure 8.2.

	A	B	C	D	E	F	G
1	Rat Maze Times						
2		Rat Name	Maze 1	Maze 2	Maze 3	Maze 4	
3		Trigger	78	103	134	146	115.25
4		Slew	86	99	145	135	116.25
5		Citation	88	112	155	154	127.25
6		Bold	81	102	168	155	126.5
7		Dancer	85	116	157	156	128.5
8							

Figure 8.1 Rat maze data including the average of all four runs.

	A	B	C	D	E	F	G
1	Rat Maze Times						
2		Rat Name	Maze 1	Maze 2	Maze 3	Maze 4	
3		Trigger	81	102	168	155	126.5
4		Slew	88	112	155	154	127.25
5		Citation	85	116	157	156	128.5
6		Bold	86	99	145	135	116.25
7		Dancer	78	103	134	146	115.25
8							

Figure 8.2 Data sorted by alphabet using column B.

Data can also be sorted by numerical values. We will sort the data by the fastest average times with the commands:

```
/Data Sort Primary-Key G7 [Enter] A Go
```

Since we had already selected a Data-Range we did not have to do that a second time. The data will be sorted using the G column and the results are shown in Figure 8.3. If ties occur in sorting the primary-key, a Secondary-Key column can also be selected to break any ties.

Plotting the Data

The data can also be plotted with these commands:

```
/Graph Type Line X C2..F2 [Enter] A C3..F3
[Enter]
        B  C4..F4  [Enter]  C  C5..F5  [Enter]
        D  C6..F6  [Enter]  E  C7..F7  [Enter]  View
```

The graph will look like Figure 8.4. Now we can add titles and a legend to the graph with the commands:

```
Other Titles First Rat Maze Timing [Enter]
  Titles Second Four Mazes [Enter]
  Titles Y-axis Time (Seconds) [Enter]
  Titles X-axis Maze Number [Enter]
  Legend A \B3 [Enter] Legend B \B4 [Enter]
  Legend C \B5 [Enter] Legend D \B6 [Enter]
  Legend E \B7 [Enter] Quit View
```

The graph will look like Figure 8.5. Note the \cell address can be used to assign the contents of a cell to appear in the legend.

	A	B	C	D	E	F	G
1	Rat Maze Times						
2		Rat Name	Maze 1	Maze 2	Maze 3	Maze 4	
3		Trigger	78	103	134	146	115.25
4		Slew	86	99	145	135	116.25
5		Bold	81	102	168	155	126.5
6		Citation	88	112	155	154	127.25
7		Dancer	85	116	157	156	128.5
8							

Figure 8.3 Data sorted using the average of all runs shown in column G.

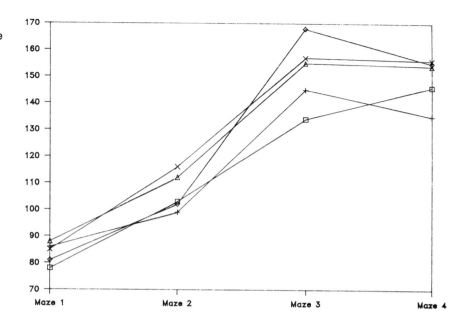

Figure 8.4 Graph showing the results of the rat maze experiments.

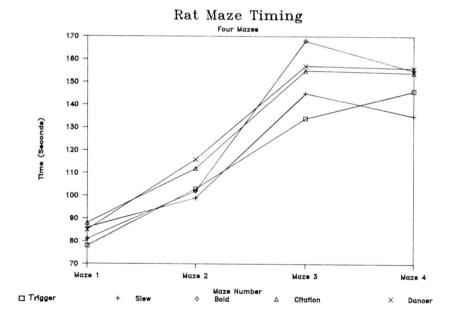

Figure 8.5 Graph with titles and legends showing results of the rat maze experiments. Applications Reporting Data #.

The Lotus data management and graphing capabilities are powerful data analysis tools. You will see many more applications of these commands in the advanced applications section.

Lotus in the Lab

9

Computing and Reporting Chromatography Results

Lotus 1-2-3 can be used for the analysis and reporting of chromatography data. Chromatographs are instruments which can separate a mixture of compounds into its pure components. This separation occurs on a chromatography column. The data generated in a chromatography experiment is a series of peaks as shown in Figure 9.1. The peaks are the response of a detector at the end of the chromatographic column. The concentration of each component in the original mixture is proportional to the area under the peak.

Two types of chromatography analysis reports will be generated: area percent, and single level standard quantitation using an external standard. The utility of having these reports performed using a Lotus spreadsheet is you can create any format you wish. What you see on the screen is the report you will get on your printer. Data in cells can be formatted to meet any reporting definition you wish or need. If additonal calculations are necessary they can be easily added to the existing report.

Once a reporting template has been created, it can be used again and again by simply entering or importing new data. The worksheet performs all of the needed calculations and a new report can be printed.

Area Percent

An area percent report computes the percentage of the total sample each component in a chromatography run represents. It assumes the detector response is identical for each component. The area percent report

Typical Chromatogram

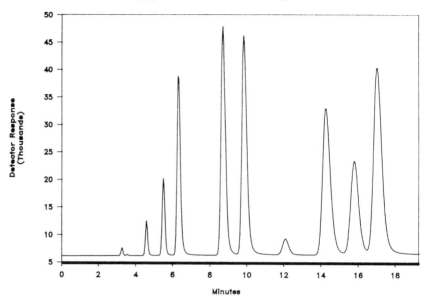

is generated by summing the areas (or heights) of each peak to compute
the total area of all peaks. Then the area of each individual peak is
divided by the total area and multiplied by 100. Figure 9.2 shows a
typical area percent report as a Lotus 1-2-3 application.

To generate this area percent report using Lotus 1-2-3 do the following.

First make sure the worksheet is empty with the command:

```
/Worksheet Erase Yes
```

You should see a blank worksheet. We will first create the the report
header information by entering the following data:

```
B1: 'Liquid Chromatography Data
C1: 'Area Percent Report
A3: \=
```

This will fill cell A3 with "========". To make this pattern across
the entire row, we can copy the contents of this cell to the other cells in
the row with the command:

```
/Copy A3 [ENTER] B3..E3 [ENTER]
```

```
           A          B          C          D          E         F
  1               Liquid Chromatography Data
  2                  Area Percent Report
  3      ===================================================
  4      Sample:   QA #2345            Date: 04-Mar-85
  5      Column:   Sephadex            Operator:GIO
  6      Solvent:  Methanol/Water      Inst No.:LC #1234
  7      Notes:    Sample vial seal broken during shipment
  8      ===================================================
  9      Peak No. Ret. Time   Area    % Total Area
 10          1        2.34   12345      1.160
 11          2        2.67   34567      3.248
 12          3        3.68   78904      7.414
 13          4        4.98    5674      0.533
 14          5        6.89   12356      1.161
 15          6        9.78  896654     84.249
 16          7       10.76    9223      0.867
 17          8       13.45   14568      1.369
 18                          ------------------
 19                         1064291   100.000
 20
```

Figure 9.2 Typical Area Percent Report generated in Lotus 1-2-3 to report chromatography data.

Now enter the header labels in these cells in columns A and D.

```
A4:  'sample:
A5:  'Column:
A6:  'solvent:
A7:  'Notes:
D4:  'Date:
D5:  'Operator:
D6:  'Inst No.:
```

Enter another "=========" pattern in row 8 by entering:

```
A8:  \=
```

Then copy the cell contents to the rest of the row with:

```
/Copy A8  [ENTER]  B8..E8  [ENTER]
```

We now have a report header. By making a few changes to this header, it could be used on many different reports. Let's save our work both because we can use this header again and also to insure we will not lose what we have done thus far. To save, enter:

```
/File Save HEADER [ENTER]
```

Now enter the column headings for the report in Row 9.

```
A9: 'Peak No.
B9: 'Ret. Time
C9: ^Area (the ^ Shift-6 will center the label in
the cell)
D9: '% Total Area
```

Enter the peak numbers, retention times and areas.

```
A10:  1    B10:  2.34    C10:  12345
A11:  2    B11:  2.67    C11:  34567
A12:  3    B12:  3.68    C12:  78904
A13:  4    B13:  4.98    C13:  5674
A14:  5    B14:  6.89    C14:  12356
A15:  6    B15:  9.78    C15:  896654
A16:  7    B16:  10.76   C16:  9223
A17:  8    B17:  13.45   C17:  14568
```

Place the underlines in cells C18 and D18 with:

```
C18: \-    D18: \-
```

Now we can do some calculations. To compute the percent of the total area of each peak, we need the total area of all of the peaks in the run. We can computer this by using the @SUM function.

```
Move to: C19
Type: +@SUM(C10..C17)
```

Note you can also interactively select the range of cells to sum by using the Lotus pointing features.

```
Move to: D10
Type: +100*C10/C19
```

This will compute the percent area of the first peak. Now we can copy this equation to the other cells in column D with:

```
/Copy D10 [ENTER] D11..D17 [ENTER]
```

Do you notice something wrong? Notice cells D11 through D17 have ERR displayed. This is the 1-2-3 error message meaning you have entered an erronous equation. Move to cell D11 and view the contents, it should be 100*C11/C20. The error is generated because cell C20 has no value and we are asking Lotus to divide by zero! But why did 1-2-3 even consider using cell C20? We wanted to use cell C19!

This can be easily explained when we review how 1-2-3 copies the contents of cells. 1-2-3 wants to help us out whenever it can. Usually if you want to copy the contents of a cell or range of cells to other cells you will need to "increment" the cell letters or numbers used in equations. In our example, the C column cell number was correctly incremented from the original C10 to C11 in the new equation. The problem, division by zero, occured because the other cell identifier C19 was also incremented to C20!

This problem is solved by using "absolute references" in a cell identifier. By placing a $ in front of a cell row or column identifier or both the row and column identifiers, it marks them as an absolute reference and indicates to 1-2-3 not to increment the cell reference when copied or moved.

Thus, the correct equation is

    D10: +100*C10/$C$19

When this equation is copied to the other cells in column D, the cell identifiers for the areas will be incremented by the total area cell, C19, will not be incremented. Copy the cell with the command:

    /Copy D10 [ENTER] D11..D17 [ENTER]

To finish this report, sum the % Total Area column with the function @SUM in D19.

    D19: +@SUM(D10..D17)

Single External Standard Quantitation

Rather than assuming the detector response is the same for each compound to be quantitated in a chromatography run, you can analyze known concentrations of standards and use the actual response to calculate the concentrations of specific components in your samples. The standards are run under the same conditions and their area or heights are recorded. Then for each compound a detector response factor is computed. The response factor is the compounds detected area (or height) divided by its concentration in the standard. To compute the concentration of a compound in an unknown sample, you simply divide the area for the compound in the sample by its response factor (Figure 9.3).

A report containing both the standard calibration data and the computed sample results is shown in Figure 9.4. To generate this report we can do the following.

We can start by making some small modifications to the Header of information we created in the Area Percent worksheet. Retrieve this worksheet with the command:

/File Retrieve **HEADER** [ENTER]

Your display will look like Figure 9.5. Now we can change the second line in the report header. First erase the contents of cell C1.

Move to: C1

```
        A          B          C          D          E          F          G
1                  Liquid Chromatography Data
2                  Single Level External Standard
3       ==================================================================
4       Sample:  QA #2345          Date:      04-Mar-85
5       Column:  Sephadex          Operator:GIO
6       Solvent: Methanol/Water    Inst No.:LC #1234
7       Notes:   Sample vial seal broken during shipment
8       ==================================================================
9       Compound Ret. Time  Area     Amount   Resp Factor   Amt %  RT Delta%
10      Methane    2.34     12345    52.630      234.56      1.317%    0.426%
11      Ethane     2.67     34567    99.936      345.89      2.501%    0.000%
12      Propane    3.68     78904   336.076      234.78      8.411%    0.271%
13      Butane     4.98      5674    23.954      236.87      0.600%    0.000%
14      Pentane    6.89     12356    50.303      245.63      1.259%    0.000%
15      Hexane     9.78    896654  3334.030      268.94     83.444%   -0.205%
16      Heptane   10.76      9223    39.319      234.57      0.984%    0.186%
17      Octane    13.45     14568    59.299      245.67      1.484%    0.148%
18                         -----------------              ---------
19                         1064291 3995.548                100.00%
20
```

Figure 9.3 Computing the concentration of a compound from its response factor. The response factor is computed from a known standard concentration of the compound.

```
                Liquid Chromatography Data
                Single Level External Standard
=================================================================
Sample:  QA #2345          Date:         04-Mar-85
Column:  Sephadex          Operator:GIO
Solvent: Methanol/Water    Inst No.:LC #1234
Notes:   Sample vial seal broken during shipment
=================================================================
Compound Ret. Time   Area     Amount    Resp Factor   Amt %   RT Delta%
Methane    2.34     12345     52.630       234.56      1.317%   0.426%
Ethane     2.67     34567     99.936       345.89      2.501%   0.000%
Propane    3.68     78904    336.076       234.78      8.411%   0.271%
Butane     4.98      5674     23.954       236.87      0.600%   0.000%
Pentane    6.89     12356     50.303       245.63      1.259%   0.000%
Hexane     9.78    896654   3334.030       268.94     83.444%  -0.205%
Heptane   10.76      9223     39.319       234.57      0.984%   0.186%
Octane    13.45     14568     59.299       245.67      1.484%   0.148%
                   ------------------                 ---------
                   1064291  3995.548                  100.00%

Calibration Data
Compound Ret. Time   Area     Amount    Resp Factor
Methane    2.35     23456       100        234.56
Ethane     2.67     34589       100        345.89
Propane    3.69     23478       100        234.78
Butane     4.98     23687       100        236.87
Pentane    6.89     24563       100        245.63
Hexane     9.76     26894       100        268.94
Heptane   10.78     23457       100        234.57
Octane    13.47     24567       100        245.67
```

Figure 9.4 Sample chromatography report containing both the standard calibration data and computed sample results.

Enter the command:

/Range Erase C1..C1 [ENTER]

Enter these cell entries and new column headings in row 9.

```
B1: 'Single Level External Standard
A9: 'Compound
B9: 'Ret. Time
C9: 'Area
D9: 'Amount
```

```
          A         B         C         D         E         F
  1              Liquid Chromatograpy Data
  2                 Area Percent Report
  3       ======================================================
  4       Sample:   QA #2345          Date:       04-Mar-85
  5       Column:   Sephadex          Operator:GIO
  6       Solvent:  Methanol/Water    Inst No.:LC #1234
  7       Notes:    Sample vial seal broken during shipment
  8       ======================================================
  9
```

Figure 9.5 Spreadsheet with header information.

```
     E9:  'Resp Factor
     F9:  'Amt %
     G9:  'RT Delta%
```

Before we complete the sample report, let's setup the calibration data section. Enter these labels and the column headings in row 23:

```
     A22:  'Calibration Data
     A23:  'Compound
     B23:  'Ret. Time
     C23:  'Area
     D23:  'Amount
     E23:  'Resp Factor
```

The calibration data can now be entered as follows:

```
     A24:  'Methane    B24: 2.35    C24: 23456    D24: 100
     A25:  'Ethane     B25: 2.67    C25: 34589    D25: 100
     A26:  'Propane    B26: 3.69    C26: 23478    D26: 100
     A27:  'Butane     B27: 4.98    C27: 23687    D27: 100
     A28:  'Pentane    B28: 6.89    C28: 24563    D28: 100
     A29:  'Hexane     B29: 9.76    C29: 26894    D29: 100
     A30:  'Heptane    B30: 10.78   C30: 23457    D30: 100
     A31:  'Octane     B31: 13.47   C31: 24567    D31: 100
```

The response factor is computed in column E by entering the formula +C24/D24, dividing the standard area by the standard amount. Now copy this formula to the other cells in column E with:

```
     /Copy E24..E24  [ENTER]  E25..E31  [ENTER]
```

Now we can complete the sample data report. Enter this data in columns A, B and C.

A10:	'Methane	B10:	2.34	C10:	12345
A11:	'Ethane	B11:	2.67	C11:	34567
A12:	'Propane	B12:	3.68	C12:	78904
A13:	'Butane	B13:	4.98	C13:	5674
A14:	'Pentane	B14:	6.89	C14:	12356
A15:	'Hexane	B15:	9.78	C15:	896654
A16:	'Heptane	B16:	10.76	C16:	9223
A17:	'Octane	B17:	13.45	C17:	14568

Now place underlines in cells C18, D18 and F18 with:

```
C18:  \-    D18:  \-    F18:  \-
```

Then sum the C, D and F columns in row 19:

```
C19:  +@SUM(C10..C17)
D19:  +@SUM(D10..D17)
F19:  +@SUM(F10..F17).
```

The other columns are generated by formulas.

```
D10:  +C10/E10
F10:  +D10/$D$19
G10:  +(B24-B10)/B24
```

This is the peaks area divided by the response factor. In cell E10 is +E24, the contents of the response factor computed from the calibration standard data. Cell F10 computes the percentage of this compound's amount of the total amounts of all compounds quantitated in the sample. Note the use of the absolute reference by using the $ signs in front of both the D and the 19. The contents of cell D19 will be used exclusively even when this formula is copied or moved. Finally, G10 is the percent difference in retention time which is computed with the formula +(B24-B10)/B24.

The formulas in row 10 can now be copied to the other rows with the copy command:

```
/Copy  D10..G10   [ENTER]   D11..G17 [ENTER]
```

The single external standard report should now appear as in Figure 9.4.

10

Fitting Curves and Plotting Functions with 1-2-3

Fitting functions to non-linear data can be performed in Lotus 1-2-3. It takes a little more work but the results are worth the effort.

When Lotus 1-2-3 version two was released one of the new menu items added was Data Regression. This menu command provides a way of performing a regression on a set of dependent (y) and independent (x) variables. The result of the regression is a set of coefficients, a_0, a_1, a_2 ... which can be used to predict the value of the dependent varible (y) when one knows the values of the independent varibles, x_1, x_2 ... . The equation has the form:

$$y = a_0 + a_1x_1 + a_2x_2 + a_3x_3 + \ldots + a_{16}x_{16}$$

Up to 16 different sets of independent variables can be handled in Lotus.

Many scientific applications can be adapted to utilize simple linear regression. Adaptations can be made to the set of values to "linearize" them. Usually this requires a simple mathematical conversion such as taking the LOG of a set of values. Conversions of this type can transform a difficult regression into a simple linear relationship. But not all phenomena can be fit to a linear regression. Higher order polynomials may be needed.

A special form of the general regression equation is:

$$y = a_0 + a_1x + a_2x^2 + a_3x^3 + \ldots + a_nx^n$$

where n can be as large as 16 and still be handled in Lotus. Here the increasing powers of x are substituted for the independent variables creating a set of polynomial equations.

This application will show how to use the Data Regression command to perform regressions on your data and plot the results. We will first build a template for performing linear regressions which includes plotting the data and the linear least squares fit line. Then we will build one to plot a quadratic (second order) fit. With this knowledge, you will be able to fit and plot equations of up to the 16th power.

Linear Least Squares Fit

The Data Regression command is very easy to use. To build our linear least squares fit template we first need some data. Let us say we are going to generate a calibration line for the analysis of methane using gas chromatography. The methane peak has a different area for different concentrations of methane in our calibration standards. We will obtain a table with concentrations and areas to which we can fit a regression line.

First clear your worksheet then enter the following set of x and y values in the A and B columns:

```
/Worksheet Erase Yes

A1: 'Calibration Data For Methane
A3: 'Concentration
B3: 'Area/1000
D3: 'LLS Value
A6: 100        B6: 210
A7: 200        B7: 304
A8: 300        B8: 400
A9: 400        B9: 489
A10: 500       B10: 610
```

Now we will execute the data regression commands on this data. The Data Regression command requires you to identify a range of cells containing the X values and the Y values. Then you can select to force the least squares fit line through zero or allow the line to have a Y-intercept. Then you identify the upper left corner of the output-range. This is where the report of the regression results will be placed on the worksheet.

Figure 10.1 Original
data with Data Regression
report for the linear fit
template.

```
            A            B         C          D          E
  1  Calibration Data For Methane
  2
  3   Concentration Area/1000           LLS Value
  4
  5
  6             100         210
  7             200         304
  8             300         400
  9             400         489
 10             500         610
 11
 12
 13
 14                   Regression Output:
 15  Constant                              107.1
 16  Std Err of Y Est                      9.672986
 17  R Squared                             0.997115
 18  No. of Observations                          5
 19  Degrees of Freedom                           3
 20
 21  X Coefficient(s)              0.985
 22  Std Err of Coef.          0.030588
 23
```

You can do all of this with the commands:

```
/ Data Regression X-Range A6..A10 [Enter]
Y-Range B6..B10 [Enter] Output-Range A14
[Enter]
Intercept Compute Go
```

The data regression report is displayed starting in cell A14. This report and the worksheet up to this point is shown in Figure 10.1. The report includes the coefficients, y-intercept and statistical estimates of potential errors in these values.

This data can be used to evaluate the regression to see if the resulting line is appropriate to use. In the next application, some of these values were used to plot confidence intervals with a linear calibration curve.

Graphing the Data and Line

Another way to evaluate the regression is to view the regression line plotted with the original data. To set up this plot, we will name a few of the cells in the regression report to make creating the linear least square fit line easy. These commands will name the cell containing the y-intercept YINT and the X coefficient or slope the name SLOPE:

```
/ Range Name Create YINT [Enter] D15 [Enter]
/ Range Name Create SLOPE [Enter] C21 [Enter]
```

To plot the linear least squares fit line we can now substitute x values into the equation y = m*x + b where m is the slope and b is the y-intercept. Enter into D6 the equation

```
D6: +$slope*A6+$yint
```

Now copy this equation to the rest of the D column. Note that by using named-ranges the contents of an equation is easy to identify.

```
/ Copy D6..D6 [Enter] D7..D10 [Enter]
```

Now let's graph the linear least squares fit line with the original data. The original data will be the A graph range and plotted with symbols only format. The linear least squares fit line will be the B graph range and plotted with LINES only format. This is done with the commands:

```
/ Graph A B6..B10 [Enter] B D6..D10 [Enter]
Options Format A Symbols B Lines Quit Quit View
```

Now let's make the graph more eye appealing by plotting the data off the side of the graph and while continuing to extend the line all the way to the edge of the plot.

Quit out of the Graph commands.

Enter these values

```
A5: 0
A11: 600
```

Then copy the least square line equation to some new cells.

```
/ Copy D10..D10 [Enter] D11..D11 [Enter]
/ Copy D6..D6 [Enter] D5..D5 [Enter]
```

Redefine the graph ranges.

```
/ Graph A {ESC} B5..B11 [Enter] B {ESC}
D5..D11 [Enter] View
```

The graph will now look like Figure 10.2 except the labels and titles will be missing.

Figure 10.2 Graph of linear least square fit with data.

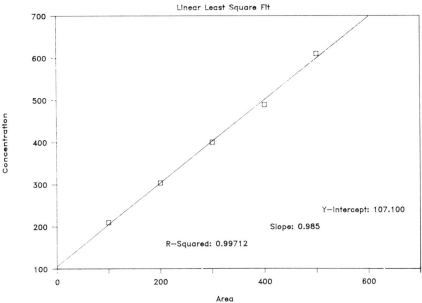

Labeling the Plot with Phantom Labels

Now we want to label the plot with the slope, Y-intercept and R Squared values for our line. To do this we will use a technique which has been called "phantom labels". Another range of cells will be defined to be plotted but rather than having symbols and lines on the graph, this range of plot cells will be used to place the labels on the graph. The location of the labels is set in the Y direction by the values in the cells. The X direction is dictated by the position of the cell in the defined plot range, i.e. all of the third data points in the plot ranges are lined up in the X direction.

First define a new plot range. We should already be in the Graph command menus.

    C E5..E11 [Enter] Quit

Now add these values in these cells

    E7: 160
    E9: 200
    E10: 240

Now use the data labels command to identify the cells with labels.

```
/ Graph Options Data-labels  C  F5..F11  [Enter]
 Right Quit Quit Quit
```

Enter the following labels in these cells

```
F7:  'R-Squared:
F9:  'Slope:
F10: 'Y-Intercept:
```

View the graph by pressing [GRAPH] (Usually the [F10] key).

Note the labels have been placed on the graph. But the symbols and lines are still being plotted. Let's get rid of the symbols and lines.

```
/ Graph Options Format  C  Neither Quit Quit Quit
```

Using Lotus String Functions

Now we want to construct the rest of our data-labels on the plot to include the actual current values for the parameters. To do this we must convert the values into labels or strings. This is done with the @STRING function. The @STRING(cell,number of decimal places) function converts a number into a label with the specified number of decimal places.

Move to H7 and enter the function

```
H7:  +@string(D17,5)
```

Now copy the label in the F column into the G column. We will join the labels in G7 and H7 with the @MID function.

```
/ Copy  F7..F7  [Enter]  G7..G7  [Enter]
```

Move to F7. Enter the joining equation

```
F7:  +@mid(G7,0,13)&@mid(h7,0,7)
```

Now let's do the same for the other values and labels. First move the labels from the F column to the G column.

```
/ Move  F9..F10  [Enter]  G9..G10  [Enter]
```

Now use the @STRING function to convert the Y-intercept and R-Squared values into labels.

```
H9:  +@string($slope,3)
H10: +@string($yint,3)
```

Now join the labels in cells F9 and F10.

```
F9: +@mid(G9,0,7)&@mid(H9,0,7)
F10: +@mid(G10,0,7)&@mid(H10,0,7)
```

Interactive Graphing Template

The template we have created can be used as a general purpose interactive graphing template. New raw data values can be entered and the resulting graph of the linear least squares fit plotted. To generate a new line do the following:

In cell B7 enter.

```
B7: 100
```

Now recalculate the data regression.

```
/ Data Regression Go
```

To view the new graph press F10 {GRAPH}.

Remember to save this worksheet with the commands:

```
/ File Save
```

Higher Order Polynomial Plots

Higher order polynomial regressions and plots can also be generated using 1-2-3. Again, let's start with a fresh worksheet and enter some data with the commands:

```
/ Worksheet Erase Yes

A1:'Quadratic Least Squares Fit Template
A4:'Y-Values
B4:'X-Values
C4:'X-Squared
A5:800          B5:20
A6:600          B6:40
A7:400          B7:60
A8:200          B8:80
A9:400          B9:100
A10:600         B10:120
A11:800         B11:140
```

Notice the y values are in the A column and the x values are in the B column in this worksheet. This worksheet will be for a second order or quadratic fit. The equation will have the form:

```
           A         B         C         D         E         F         G         H
 1    Quadratic Least Squares Fit Template
 2
 3
 4    Y-Values  X-Values  X-Squared                   Regression Output:
 5       800       20        400       Constant                         1228.571
 6       600       40       1600       Std Err of Y Est                 75.59289
 7       400       60       3600       R Squared                        0.923076
 8       200       80       6400       No. of Observations                     7
 9       400      100      10000       Degrees of Freedom                      4
10       600      120      14400
11       800      140      19600       X Coefficient(s)   -22.8571 0.142857
12                                     Std Err of Coef.   3.375582 0.020619
13
14
```

Figure 10.3 Original data with Data Regression report for the quadratic fit template.

$$y = Ax^2 + Bx + C$$

The C column will be used to hold the squared values of x. In C5 enter the equation: +B5^2

Then copy this equation to the other C column cells with the command:

```
/ Copy  C5..C5  [Enter]  C6..C11  [Enter]
```

Now we can execute the data regression command again. This time, the X-Range will have two columns of values representing values of x and x squared. Use the command:

```
/ Data Regression  X-Range B5..C11  [Enter]
Y-Range A5..A11 [Enter]  Output-Range E4
[Enter]
Intercept Compute Go
```

The regression report and worksheet should look as shown in Figure 10.3.

Notice the Coefficient(s) part of the report now has two sets of values. The first set is the x coefficient and the second set is the x^2 coefficient.

Plotting the Quadratic Equation

In our linear least squares fit example, it was easy to draw the line by selecting just a few points and connecting them with a straight line. To plot the curve in a quadratic equation more data points must be plotted. Even though each line between points is a straight line, your eye will

Table 10.1

	Pixels in X direction
CGA	320
EGA	640
Hercules	720
VGA	640
Plotter	900 or more

perceive the lines as a curve if there are enough points and they are narrowly spaced.

The question becomes how many points should be plotted. Normally this can be answered by considering the resolution of the graphics device you will be using to display the curve. The number of points will be different for a medium resolution monitor versus a high resolution monitor or plotter. A good rule of thumb is to use the same number of data points as there are pixels of resolution in the x direction. Refer to Table 2 for various resolutions of monitors and plotters. Always be ready to experiment with the number of points to plot. Try out a few plots with different numbers of points and see what looks best. This is especially important if your ultimate output device is a plotter.

Plotting the Curve

The curve can be plotted by first entering a column of x values over the range of values you wish to see plotted. This can be most easily done with the Data Fill command. This command allows us to fill worksheet cells with values. Let us view 360 points in the x value range from 0 to 180. Use the command:

```
/ Data Fill E14..E374 [Enter] 0 [Enter] .5
[Enter] 180 [Enter]
```

It is easier to create formulas if you first name cells with the Range Name command. Then you can use the more meaningful name in the formula. First we name the cells where our regression results are displayed:

```
/ Range Name Create YINT [Enter] H5 [Enter]
/ Range Name Create XCOEF [Enter] G11 [Enter]
/ Range Name Create X2COEF [Enter] G12
[Enter]
```

Now enter a formula for the y value. Note you must use the absolute reference for each of the regression results. This will allow us to copy the formula to the other cells in the F column correctly.

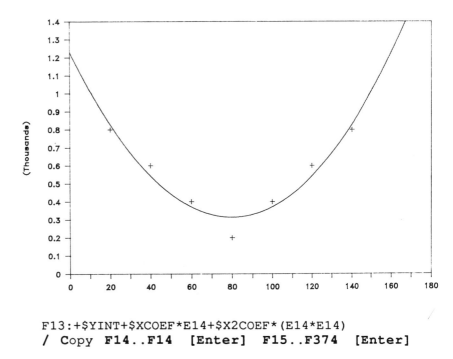

Figure 10.4 Quadratic equation regression curve plotted with original data.

```
F13:+$YINT+$XCOEF*E14+$X2COEF*(E14*E14)
/ Copy F14..F14  [Enter]  F15..F374  [Enter]
```

To include the original data in the plot copy the x and y values to locations where they can be plotted. The x values will be placed just below the large column of X values with one blank cell separating the two sets of data. The y values will be placed in the G column. A second graph range will be used to plot this data so the symbols only format can be used to plot this data.

```
/ Copy B5..B11  [Enter]  E376..E382  [Enter]
/ Copy A5..A11  [Enter]  G376..G382  [Enter]
```

Now plot the data with these graph commands:

```
/ Graph Type XY X E14..E382 [Enter] A
F14..F382  [Enter]
B G14..G382 [Enter] Options Format A Lines B
Symbols Scale X
Manual Lower 0 [Enter] Upper 180 [Enter]
Y Manual Lower 0 [Enter] Upper 1400 [Enter]
Quit Quit View
```

The resulting plot should look like Figure 10.4. Parameter values could be added to this plot using the same "phantom label" techniques described earlier.

Figure 10.5 Cubic
equation regression curve
plotted with original data.

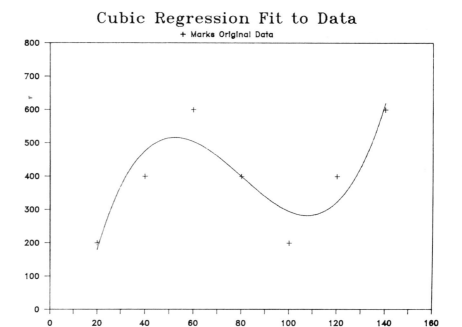

Cubic Regression Fit to Data

Using the same types of commands you should be able to construct a cubic fit and plot like the one shown in Figure 10.5. Other types of fits and plots can also be created in Lotus 1-2-3.

11

Geometric Regression and Plotting with Lotus 1-2-3

Linear and polynomial regressions are not the only kinds of fits you can perform on your data using 1-2-3. Other fits require a little equation set up, but with a little work they can be done. In 1-2-3 you can compute the regression values by entering the complete equations to compute the needed coefficients. You can then substitute in specific independent variable values to find the values of the corresponding dependent variables.

Linear and polynomial regressions can take you a long way toward fitting your data to mathematical equations for predicting the outcome of an analysis. (See the previous application for templates which perform linear and polynomial regressions). But some of the systems we would like to monitor do not follow simple linear or polynomial rules. Some equations can be linearized with simple transformations. See the following application on Transforming Data to get Linear Results. In that application we will solve the same problem as in this article, except we will transform the regression data and then use the built in commands for Data Regression. Other sets of data can not be fit without more exotic correlation of the data. In those cases, the solutions must be created using brute mathematical force.

These more exotic fits can also be performed using Lotus 1-2-3. Instead of relying on the Data Regression commands, you can build your own regression equations. Then you can fit your data and plot the regression curve. This application will describe how to create a template to make power curves or geometric fits to a set of data.

Power Curves

Power curves or geometric equations follow the general form:

$$y = ax^b$$

where

> y is the dependent variable
> x is the independent variable
> a and b are constants

The values of a and b can be computed from a set of x and y data using these equations:

```
b=(log xi)(log yi) -((log xi)*(log yi))/n
(log xi)2-((log xi)2)/n
a=10(log yi)/n-b*(log xi)/n
```

where

> n = the number of x,y pairs used for the regression

Power Curve Application

Power curves are used to correlate gas volume and pressure data. If you make measurements of the pressure and volume of a gas they will follow the general geometric form:

$$p = av^{-b}$$

where

> p is pressure
> v is volume
> a and b are constants, b is called the polytropic constant.

Let us set up a template which fits a power curve to the following data and calculates the values for the constants a and b. Using the constants we will compute the pressure with a volume of 20. Finally, we will set up the template to plot the resulting power curve and data points.

volume	pressure
10	210
15	108
30	40
50	12
70	9
90	6.8

The Template

The worksheet we are going to create is shown in Figure 11.1. To build this template first initialize your 1-2-3 worksheet and then enter these labels:

```
/Worksheet Erase Yes
A1: 'Power Curve Fit y=ax^b
D1: 'points =
D2: 'a =
F2: 'b =
A4: 'x-values
B4: 'y-values
D4: 'log(x)
E4: 'log(y)
F4: 'lgx * lgx
G4: 'log(x)^2
H4: 'log(y)^2
```

Then enter the x and y values into columns A and B starting in cell A5.

```
A5:  10        B5:  210
A6:  15        B6:  108
A7:  30        B7:  40
A8:  50        B8:  12
A9:  70        B9:  9
A10: 90        B10: 6.8
```

Setting-up Intermediate Values

Since there are so many values packed into the equations, a good strategy is to break the large equation into a number of intermediate values. Then the intermediate values can be combined for the final regression value. Five sets of intermediate values are constructed. In the D column we place the Log of the x or volume values. Similarly we place the Log of the y or pressure values in column E, the product of Log x and Log y in column F, Log x squared in column G and finally Log y squared in column H. This can be done for our template with the following commands:

```
D5: +@log(a5)
E5: +@log(b5)
F5: +d5*e5
G5: +d5*d5
H5: +e5*e5
/Copy D4..H5  [Enter]  D6..H10  [Enter]
```

By naming ranges, it will make constructing the equations much easier. Also by using named ranges, when we want to expand the number of data points we wish to use for the fit, we simply change the ranges for each name rather than having to rewrite or edit the large equations. Name the necessary ranges with the commands:

	A	B	C	D	E	F	G	H
1	Power Curve Fit y=ax^b			points =	6			
2				a = 8574.692		b = -1.61455		
3								
4	x-values	y-values		log(x)	log(y)	lgx*lgy	log(x)^2	log(y)^2
5	10	210		1	2.322219	2.322219	1	5.392702
6	15	108		1.176091	2.033423	2.391491	1.383190	4.134812
7	30	40		1.477121	1.602059	2.366436	2.181887	2.566596
8	50	12		1.698970	1.079181	1.833496	2.886499	1.164632
9	70	9		1.845098	0.954242	1.760670	3.404386	0.910578
10	90	6.8		1.954242	0.832508	1.626924	3.819063	0.693071
11								

Figure 11.1 Data for power curve.

```
/ Range Name Create lgx [Enter] D5..D10
[Enter]
/ Range Name Create lgy [Enter] E5..E10
[Enter]
/ Range Name Create lgxlgy [Enter] F5..F10
[Enter]
/ Range Name Create lgx2 [Enter] G5..G10
[Enter]
/ Range Name Create lgy2 [Enter] H5..H10
[Enter]
```

Now complete the equations and compute the results.

```
E1: 6
E2: +10^((@SUM(lgy)/E1)-(G2*(@SUM(lgx)/E1)))
G2: +(+E1*@SUM(lgxlgy)-(@SUM(lgy)*@SUM(lgx)))/
    ((+E1*@SUM(lgx2)-(@SUM(lgx))^2)
```

After entering these two equations your template should look like Figure 11.1 except for the plotting values starting in cell A5 and C5. Your values should be exactly like the ones shown in Figure 11.1. If your values are not the same, you have probably entered the equations wrong. Check the equations you have entered, especially check the number of parentheses you have entered.

Predicting Results

From the constants we have computed, we can now compute the pressure when the volume is 20. We do this by simply substituting in the values for each variable, a, b, and v.

```
E12: +E3*(20^G2) [Enter]
```

The pressure value, 68.01874, for a volume of 20 will appear in cell E12.

Plotting the Data and Curve

A plot of the power curve can be generated along with the original data points used to compute the regression. To make the plot, we will compute the pressure at a number of volume values. Then we will plot the data. To do this enter the following commands:

```
/Data Fill A12..A60 4 [Enter] 2 [Enter] 100
[Enter] C12: +$E$2*(A12^$G$2)
```

Be sure you use the absolute reference identifier ($) for the $E$2 and $G$2 cells.

```
/Copy C12..C12 [Enter] C13..C60 [Enter]
```

Now we can plot this graph and the original data with these commands:

```
/Graph Type XY X A5..A60 A B5..B60 B C5..C60
Options
Format A Symbols B Lines Quit
Titles First Power Curve Fit [Enter]
Second y = a * x^b [Enter] Y-Axis pressure
[Enter]
X-Axis volume [Enter] Quit View
```

The graph will look like figure 11.2.

Many Other Fits Possible

After working through this example, you should be convinced that you can construct nearly any type of regression using Lotus 1-2-3. The resulting curve or line can also be plotted using Lotus graphics and results predicted by substituting in the appropriate values using the Lotus spreadsheet.

Figure 11.2 Graph of geometric fit with original data.

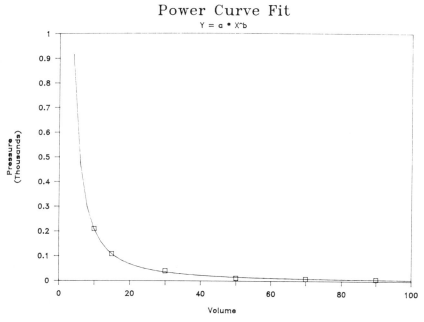

12

Plotting Confidence Intervals

Once a linear regression line has been fit to a set of calibration data, the next step is to use the line to predict the concentrations of samples from their observed experimental readings (areas, heights, ODs). We need to know how good our predictions will be. Often these predictions will be given in the form of an interval together with a confidence coefficient associated with the interval, i.e., a confidence interval estimate.

The best way to visualize these confidence intervals is to plot them along with the regression line. This template will generate a plot of five data points with a confidence interval for 95 percent correct predictions.

Enter in the raw data for this application:

```
B1: 'Conc        C1: 'Area
B3: 100          C3: 203450
B4: 200          C4: 250000
B5: 300          C5: 604908
B6: 400          C6: 827980
B7: 500          C7: 1030890
```

We will first perform a linear regression on this data with the following commands:

Figure 12.1 Regression output for the confidence interval application.

	A	B	C	D	E	F
1		Conc	Area			
2						
3		100	203450			
4		200	250000			
5		300	604908			
6		400	827980			
7		500	1030890			
8						
9						
10						
11						
12						
13		Regression Output:				
14		Constant			−86412.4	
15		Std Err of Y Est			76333.57	
16		R Squared			0.966126	
17		No. of Observations			5	
18		Degrees of Freedom			3	
19						
20		X Coefficient(s)	2232.86			
21		Std Err of Coef.	241.3879			
22						

```
/ Data Regression X-Range B3..B7 [Enter]
Y-Range C3..C7 [Enter] Output-Range B13
[Enter]
Intercept Compute Go
```

The display should look like Figure 12.1, with the regression output printed below the data.

Now we will name a few cells and set up the regression line with the commands:

```
/ Range Name Create YINT [Enter] E14 [Enter]
/ Range Name Create SLOPE [Enter] D20 [Enter]
```

Enter in cell D1 the label: 'slope*x+yint

Enter in cell D3 the equation: +$slope*b3+$yint

Be sure you use the absolute reference symbols ($) in the equation. Now copy the equation to the other cells in the D column with the command:

```
/Copy D3 [Enter] D4..D7 [Enter]
```

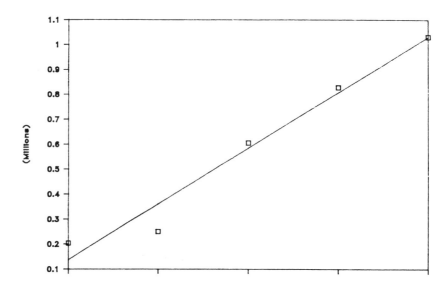

Figure 12.2 Crude calibration curve graph.

Plotting the Graph

We can plot this data with the following commands:

```
/Graph A C3..C7 [Enter] B D3..D7 [Enter]
Type Line Options Format A Symbols B Lines Quit
Quit View
```

The crude graph will look like Figure 12.2.

Lets place the data points so they are off the margins of the graph and continue to extend the line across the graph. To do this we will define the A plot range to have a blank cell at the front and end, and add new points to the B plot range.

Quit out of the Graph command menu.

First enter in cell B2 the value 0. Then in cell B7 enter the value 600. Now copy the equation in D3 to cells D2 and D7 with:

```
/Copy D3 [Enter] D2..D7 [Enter]
/Copy D3..D3 to D7..D7
```

Now extend the A and B plot ranges with:

```
/Graph A [ESC] C2..C8 [Enter] B [ESC] D2..D8
[Enter]
X B2..B8 [Enter] View Quit
```

The graph should look like Figure 12.3.

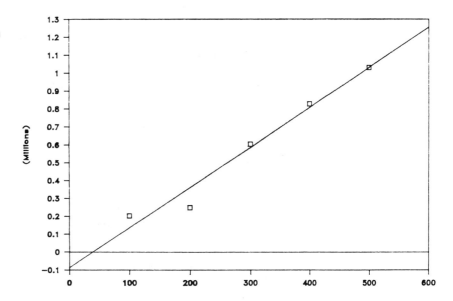

Figure 12.3 Linear regression plot of calibration data.

The Confidence Bands

Now we need to add the confidence bands. The confidence band values (BV) at each data point are computed from our data and added and subtracted from the regression line value to form the confidence bands. The equation for the band value is:

```
BV = SQRT(2*F)*YERROR*SQRT((1/N)+((X-X)2/Sxx))
```

where

F = F distribution value from the proper F-table
$YERROR$ = computed error in the Y intercept value
N = Number of observations
X = Data Reading X = Average of all data readings
$Sxx = \sum x^2 - (\sum x)^2/N$

To set up our template to make these calculations first we will perform some intermediate calculations.

In cell A9 enter the label: Intermediate Results for Confidence Band Calculations

Name some more cells with named ranges with the commands:

```
/ Range Name Create AVG [Enter] B12 [Enter]
/ Range Name Create FVAL [Enter] B11 [Enter]
/ Range Name Create NUMPTS [Enter] E17
[Enter]
```

```
/ Range Name Create SXX [Enter] E11 [Enter]
/ Range Name Create YERROR [Enter] E15
[Enter]
```

Then enter these labels and equations in the following cells:

```
A10: "Sum:      A11: "Fval: A12: Avg:
B10: +@SUM(B3..B7)
B11: 5.79   (F value for two degrees of freedom and
5 readings)
B12: +@AVG(B3..B7)
```

The F value is read from a common statistics table listing F values. The table values are selected by degrees of freedom and number of observations. A part of this table is shown in Figure 12.4.

Compute the product of each X value in the E column by first entering in cell E3 the equation:

```
E3: +B3*B3
```

Then copy this equation to the rest of the column with:

```
/Copy E3 [Enter] E4..E7 [Enter]
```

In the F column compute the X values minus the average X value. In cell F2 enter the equation:

```
F2: +B2-$AVG
```

Remember to use the absolute reference for the average value. Then copy this equation to the rest of the column with:

```
/Copy F2 [Enter] F3..F8 [Enter]
```

Figure 12.4 A portion of the F Value table.

Observations	Degrees of Freedom-->					
	1	2	3	4	5	6
1	161.44	200	216	225	230	234
2	18.51	19	19.2	19.2	19.3	19.3
3	10.13	9.55	9.28	9.12	9.01	8.94
4	7.71	6.94	6.59	6.39	6.26	6.16
5	6.61	5.79	5.41	5.19	5.05	4.95
6	5.99	5.14	4.76	4.53	4.39	4.28
7	5.59	4.74	4.35	4.12	3.97	3.87
8	5.32	4.46	4.07	3.84	3.69	3.58

Now complete the intermediate results with these equations:

```
D10: +B10*B10/numpts
E10: +@SUM(E3..E7)
E11: +E10-D10
```

The Band Value

Now enter the equation for the band value in cell G2:

```
G2:+@SQRT(2*$FVAL)*$YERROR*@SQRT((1/$NUMPTS)+((F2*F
2)/$SXX))
```

Again, be certain you enter all of the absolute reference symbols and spell the named ranges correctly.

You can then copy this equation to the rest of the column with:

```
/Copy G2 [Enter] G3..G8 [Enter]
```

Columns H and I are the regression values with the band value added and subtracted.

In cell H2 enter the equation:

```
H2: +D2+G2
```

In cell I2 enter the equation:

```
I2: +D2-G2
```

Copy these cells to the rest of the column with the command:

```
/Copy H2..I2 [Enter] H3..I8 [Enter]
```

Plotting the Bands

Plotting the confidence bands is easy with the following commands:

```
/Graph C H2..H8 [Enter] D I2..I8 [Enter]
    Options Format C Lines D Lines Quit Quit View
```

The graph will be as shown in Figure 12.5.

Adding Parameter Labels

Only the parameter values are missing from the plot. These are added by defining another range of cells to plot and using the "neither" format so you actually do not see the plot symbols or lines. Data-labels are assigned to the locations and are plotted on the graph.

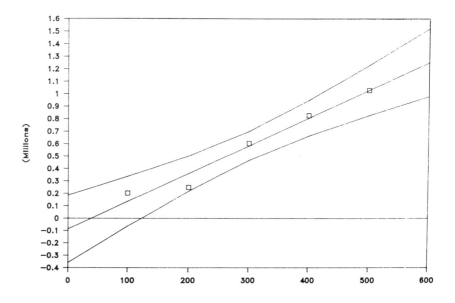

Figure 12.5 Graph with confidence bands. Lacks only the slope, y-intercept and R Squared values.

Quit out of the Graph command menu.

Start by entering the following values in the cells:

```
F13: +C3
F14: 300000
F15: 400000
```

These values will place the labels in the Y direction.

Enter the data labels in the G column:

```
G13: +@MID("Y-Int:
",0,7)&@MID(@STRING(YINT,1),0,10)
G14: +@MID("R Squared:
",0,11)&@MID(@STRING(E16,4),0,10)
G15: +@MID("Slope:
",0,7)&@MID(@STRING(SLOPE,2),0,10)
```

These commands convert the values in the various cells into strings of text using the @STRING function. The @STRING function allows you to convert a value into text with a specified number of decimal places. The text is then combined with a label using the @MID function. These text strings can then be placed on the graph in various positions as data-labels.

Figure 12.6 Final graph for the confidence band template.

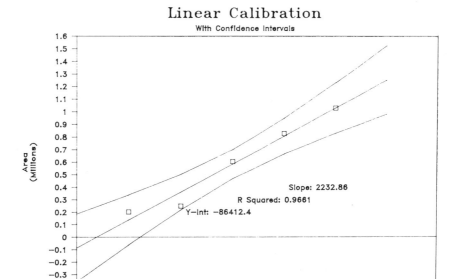

Linear Calibration
With Confidence Intervals

Now we define a new graph plot range to set the labels in the X direction.

```
/Graph E F11..F18 [Enter] Options Format E
Neither Quit
Data-Labels E G11..G18 [Enter] Right Quit Quit
View
```

The graph looks like the one in Figure 12.6. This graph shows the regression line with confidence bands.

13

3D Plotting With Lotus 1-2-3

Lotus add-in programs provide many new capabilities to the popular spreadsheet program.

Lotus Development has opened up Lotus 1-2-3 by providing the information and programming tools to allow third party developers to add new functions and capabilities to the basic spreadsheet program. These "add-in" programs are loaded right into memory with Lotus 1-2-3 and have full access to the data in the worksheet. A number of different programs are already available including wordprocessors, data managers, and graphics programs. This application will describe how a 3D graphing add-in program, called 3D Graphics, can be used within 1-2-3 to generate many new graphs from your data.

Beyond Simple Graphics

Lotus 1-2-3 can create many different types of graphs. Most of them are very basic line, bar and pie graphs. Although you can generate many different types of plots using the "blank cell" techniques described in other applications in this book, it is nice to have new plots which can be generated by just selecting them from a menu.

3D plotting is an excellent capability provided by an add-in program. To attempt to perform 3D plotting with the basic Lotus graph commands would be a very difficult task. The add-in program can use the data you have entered into your Lotus worksheet and create many additional plots. The graphs you create are saved in regular "PIC" files can can be plotted using the same PGRAPH program you would use to plot regular Lotus graphics.

Installing a Lotus Add-in

3D-Graphics from Intex Solutions Inc. gives you the ability to see your Lotus 1-2-3 data plotted in three dimensions. To use the program is as easy as selecting commands off a typical Lotus menu. You first must install the Lotus Add-in Driver Manager. Then you attach the 3D add-in, which means the program is loaded into 1-2-3's memory area. Then you can call up the add-in with a special keystroke and run the program right while you are running 1-2-3. The commands for the add-in use the same menu structures as the regular 1-2-3 commands.

The Add-in Driver Manager (ADD_MGR.EXE) is a special program written by Lotus Development which allows the use of add-in programs. This program lets you attach, detach and invoke add-ins. The program is provided on the 3D-Graphics release disk. The Add-in Driver Manager is added to the Lotus driver "set" file with the command:

```
ADD_MGR 123.SET
```

The default driver set file is named 123.SET. If you want to have a separate driver set file you can enter the command:

```
ADD_MGR driver.set.name
```

Where driver.set.name is the name of the driver set you want to use.

Attaching the Add-in

After you have added the Add-in Manager to your driver set file you can run Lotus 1-2-3. You must make sure the file 123.DYN is present in the 1-2-3 default directory. To attach 3D-Graphics first make sure you are in the READY mode. Then you press the ALT-F10 keystroke combination and the Add-in Manager menu will appear:

```
A1:
Attach  Detach  Invoke  Clear  Setup  Quit
Attach  an  application
        A          B        C        D
    1
    2
    3
```

You then choose the Attach option and select the .ADN file for the display you are using. As you can see, 3D supports the color graphics adapter (CGA), enhanced graphics adapter (EGA) and Hercules monochrome graphics (HERC).

```
A1:
Enter  application  to  attach:  C:\123\*.ADN
3DG_CGA.ADN          3DG_EGA.ADN                3DG_HERC.ADN
        A            B         C          D
1
2
3
```

You can then select a keystroke combination which will invoke 3D-Graphics from within 1-2-3. Here ALT-F9 is selected as the invoking keystroke:

```
A1:
No key      7       8       9
Alt-9
        A           B           C           D
  1  2  3
```

When you have successfully attached the 3D-Graphics program, the program's logo will appear on the screen.

Starting
3D-Graphics

Let's plot some data we have collected from a chromatography analysis on a reaction we are trying to optimize. The data is shown in Figure 13.1. We have collected data on a reaction where the concentration of an expensive precursor was changed from 1.1 to 1.8 millimoles per liter. The effluent from our preparative liquid chromatograph was collected in different tubes. Each tube was then chromatographed a second time to find the concentration of our needed compound. The concentration of the compound in tubes 8 through 14 are shown in the table.

To start 3D-Graphics

```
Press: [ALT]-[F9]
```

You will then see the 3D-Graphics menu:

```
A1:
Type    X    Y    A    View    Reset    Save    Options
Display      Name      Quit
        A              B           C           D
E           F
1
2
3
```

	A	B	C	D	E	F	G	H
1	Synthesis of Compound XX-678		Optimization of Reactants Study					
2		Concentration of XX-678 found via Liquid Chromatography						
3				Collection Tube Numbers				
4	Conc of							
5	precursor	8	9	10	11	12	13	14
6	XY-P678							
7	1.1	14.7	102.9	120.4	90.3	53.4	17.6	1.4
8	1.2	20.4	132.1	148.3	103.9	57.4	17	1.3
9	1.3	22.2	132.4	138	87.2	44.6	12.9	0.9
10	1.4	30.8	156.8	160.9	93.6	45.6	12.5	0.8
11	1.5	40.9	178.4	182.1	104.7	48.6	13.4	0.9
12	1.6	48.9	162.8	162.7	88.7	40.3	10.3	0.7
13	1.7	43.8	132.7	135.2	65.3	25.5	6.3	0.4
14	1.8	30.3	108.8	124.5	63.9	20	4.2	0.4
15								

Figure 13.1 Data for 3D plotting. Any data which you would like to view in 3 dimensions can be used. In this example, the collection tubes are the X-axis, the concentration of precursor XY-P678 is the Y-axis and the concentration of the compound XX-678 in each tube is the Z-axis.

Figure 13.2 3D Bar graph of the data shown in Figure 13.1.

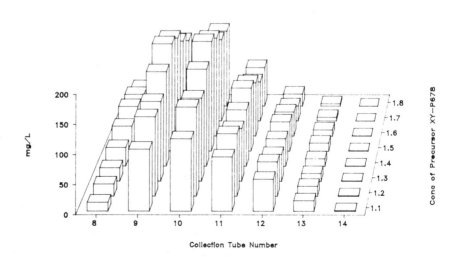

Synthesis of XX-678 Optimization Study

Conc found via Liquid Chromatography

Figure 13.3 3D Surface plot of the data shown in Figure 13.1. Colors can be used in these plots if you have a color monitor and plotter. Color contours, up to 16, can be added to the plot.

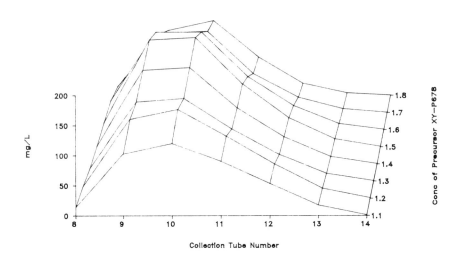

Synthesis of XX−678 Optimization Study

Conc found via Liquid Chromatography

We then select A from the 3D-Graph menu and choose the range B7..H14 for our data to plot. Then we chose the graph type, and label the axes with the commands:

```
Type Bar
Options Titles First
  Synthesis  of  XX-678  Optimization  Study
[Enter]
Second  Conc  found  via  Liquid  Chromatography
[Enter]
Z-axis  mg/L  [Enter]
Y-axis  Conc  of  Precursor  XY-P678  [Enter]
X-axis  Collection  Tube  Number  [Enter]
Quit
```

Then select View from the menu and you will see the graph shown in Figure 13.2.

Surface graphs and line graphs can also be generated. A surface graph of our data is shown in Figure 13.3.

Changing Viewing Points

You can change the viewing direction and view angle. Viewing direction can be changed in 90 degree increments (90, 180, and 270). Figure 13.4 shows the data being viewed from 90 degrees from our original viewing point. The angle of the view can also be selected. Only three angles are provided: low, medium, and high.

Figure 13.4 View
direction and angle can be
changed on the 3D plots.
View directions are in
increments of 90 degrees
(0, 90, 180, 270) while
three view angles (low,
medium and high) are
available.

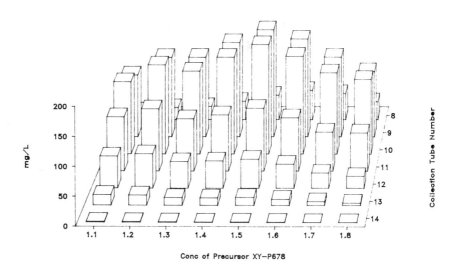

Synthesis of XX−678 Optimization Study
Conc found via Liquid Chromatography

Limitations

The only limitation found with 3D-Graph is the maximum range you
can plot is a 100 x 100 data point range. This is more than sufficient for
most laboratory applications except for viewing raw data. Multiple
raw data sets would be difficult to squeeze into memory anyway, so the
size limitation is not critical. Since 3D-Graphics creates standard Lotus
PIC files for output to printers and plotters, any other software you are
currently using to enhance your graphics, like Lotus Freelance, can still
be used. If you have data you would like to analyze or display in 3D,
this program should be one of the first Lotus add-in programs you
should have in your Lotus library of add-in software.

3D Graphics
A Lotus 1-2-3 Add-in program for generating three-dimensional graphs.
Requires 1-2-3 Release 2.0 or greater.
3D-Graphics requires 55K of standard memory.
Price: $79.95

Intex Solutions Inc.
568 Washington St.
Wellesley, MA 02181
(617) 239-1168

14

Semi-Log Plots for Data Analysis

A common method used in computerized data analysis is to transform observed or measurable values using a mathematical function or equation. The transformed values can then be viewed to follow a simple trend. Usually the trend is a straight line. By performing the transformation a complex set of values becomes easy to understand. The transformation becomes a computable model for the observations and can be used to predict the results for measurements or for calibration.

For example, in gel permeation chromatography (GPC) for certain molecular weight ranges, the retention times for polymers are linear with the log of their molecular weights. A typical set of calibration data for GPC analysis is shown in Figure 14.1 plotted linear X and Y axes. The data presented in this form seems to follow no recognizable trend. But when the molecular weight axis is changed to a log scale as shown in Figure 14.2, the data can be easily observed to follow a straight line.

Lotus 1-2-3 is well suited to perform this type of mathematical transformation and graphing. You can build nearly any mathematical transformation with the math operators and functions found in 1-2-3. The results of the transformation can then be plotted using graphics. To provide concrete examples of performing this type of data analysis we will build a semi-log plotting template. By simply changing the equations, the same template can be used to plot log-log and other functions.

Figure 14.1 Typical GPC calibration data showing the retention time and molecular weights of the polymers. This plot including the "table" was made in Lotus 1-2-3.

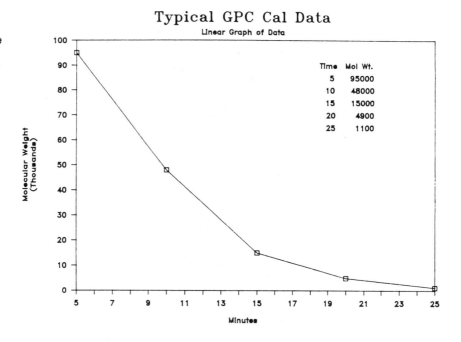

Figure 14.2 The same data shown in Figure 14.1 is now plotted after transforming the molecular weight scale to the Log of the molecular weight. The results now follow a linear trend.

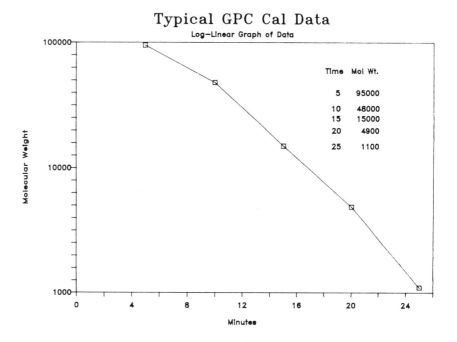

Even though Lotus 1-2-3 can only use linear scales for plotting it is possible to plot other functions and use other scales. You will need to use two new techniques, normalization of the transformed values and use the graph data-labels to re-label the axis.

Normalization

Normalization of the transformed values creates a new set of numbers which fall between zero and one. Transformation of the observed values and normalization can be performed in the same equation. To normalize a value you take the transformed value, subtract the lowest value to display, and then divide by the highest value minus the lowest value. For example when you use log to transform a value the normalization would be:

$$\frac{(@LOG(value) - @LOG(lowest\ value))}{(@LOG(highest\ value) - @LOG(lowest\ value))}$$

This equation creates values which fall between zero and one that will display the characteristics of the transformed values.

Setting Up the Linear Section

Lets create our semi-log plot template shown in Figure 14.3. Start with a clean worksheet. The A column will hold the observed values and the B column the transformed and normalized values. Use the Data Fill command to enter the values in the A column. Place the cursor in cell A2 and enter:

```
/Data Fill A2..A27 [Enter] 0 [Enter] 1
[Enter] 25 [Enter]
```

Now in B2 enter the equation:

```
B2:+(A2-$A$2)/($A$27-$A$2)
```

Be sure you use the absolute reference designations for the highest ($A$20) and lowest ($A$2) values. By using the absolute reference we can now copy this equation to the other B column cells and the equation will use the correct cells. Copy the equation to the other cells:

```
/Copy B2 [Enter] B3..B27 [Enter]
```

Setting up the Log Section

Now move to cell A40. Use the Data Fill command to fill in the values in the A column:

```
/Data Fill A40..A49 [Enter] 1000 [Enter]
  1000 [Enter] 10000 [Enter]
/Data Fill A50..A58 [Enter] 20000 [Enter]
  10000 [Enter] 100000 [Enter]
```

Figure 14.3

	A	B	C	D	E	F	G	H	I
1	Time	Transform	0	0		0		Time	Mol Wt.
2	0	0	0	1				5	95000
3	1	0.04						10	48000
4	2	0.08	0.04	0				15	15000
5	3	0.12	0.04	1				20	4900
6	4	0.16						25	1100
7	5	0.2	0.08	0					
8	6	0.24	0.08	1					
9	7	0.28							
10	8	0.32	0.12	0					
11	9	0.36	0.12	1					
12	10	0.4							
13	11	0.44	0.16	0					
14	12	0.48	0.16	1					
15	13	0.52							
16	14	0.56	0.2	0		5			
17	15	0.6	0.2	1					
18	16	0.64							
19	17	0.68	0.24	0					
20	18	0.72	0.24	1					
21	19	0.76							
22	20	0.8	0.28	0					
23	21	0.84	0.28	1					
24	22	0.88							
25	23	0.92	0.32	0					
26	24	0.96	0.32	1					
27	25	1							
28			0.36	0					
29			0.36	1					
30									
31			0.4	0		10			
32			0.4	1					
33									
34			0.44	0					
35			0.44	1					
36									
37			0.48	0					

(continued)

Figure 14.3 (continued)

	Mol Wt.	Transform			
38			0.48	1	
39	Mol Wt.	Transform			
40	1000	0	0.52	0	
41	2000	0.150514	0.52	1	
42	3000	0.238560			
43	4000	0.301029	0.56	0	
44	5000	0.349485	0.56	1	
45	6000	0.389075			
46	7000	0.422549	0.6	0	15
47	8000	0.451544	0.6	1	
48	9000	0.477121			
49	10000	0.5	0.64	0	
50	20000	0.650514	0.64	1	
51	30000	0.738560			
52	40000	0.801029	0.68	0	
53	50000	0.849485	0.68	1	
54	60000	0.889075			
55	70000	0.922549	0.72	0	
56	80000	0.951544	0.72	1	
57	90000	0.977121			
58	100000	1	0.76	0	
59			0.76	1	
60					
61			0.8	0	20
62			0.8	1	
63					
64			0.84	0	
65			0.84	1	
66					
67			0.88	0	
68			0.88	1	
69					
70			0.92	0	
71			0.92	1	
72					
73			0.96	0	
74			0.96	1	
75					
76			1	0	25

77	1	1	
78			
79			
80			
81			
82			
83			
84			
85	0	0	1000
86	1	0	
87			
88	0	0.15051	2000
89	1	0.15051	
90			
91	0	0.23856	3000
92	1	0.23856	
93			
94	0	0.30102	
95	1	0.30102	
96			
97	0	0.34948	5000
98	1	0.34948	
99			
100	0	0.38907	
101	1	0.38907	
102			
103	0	0.42254	
104	1	0.42254	
105			
106	0	0.45154	
107	1	0.45154	
108			
109	0	0.47712	
110	1	0.47712	
111			
112	0	0.5	10000
113	1	0.5	
114			
115	0	0.65051	20000

(continued)

Figure 14.3 (continued)

116	1	0.65051	
117			
118	0	0.73856	30000
119	1	0.73856	
120			
121	0	0.80102	
122	1	0.80102	
123			
124	0	0.84948	50000
125	1	0.84948	
126			
127	0	0.88907	
128	1	0.88907	
129			
130	0	0.92254	
131	1	0.92254	
132			
133	0	0.95154	
134	1	0.95154	
135			
136	0	0.97712	
137	1	0.97712	
138			
139	0	1	100000
140	1	1	
141			
142	0.2	0.98886	
143	0.4	0.84062	
144	0.6	0.58804	
145	0.8	0.34509	
146	1	0.02069	

Enter in B40 the equation:

```
B40: +(@LOG(A40)-@LOG($A$40))/(@LOG($A$58)-
@LOG($A$40))
```

and copy it to the other cells.

```
/Copy B40 [Enter] B41..B58 [Enter]
```

Using XY Graph

The graph we will generate uses the XY Graph Type. The graph is defined by two columns of values. One column for the X values and one for the Y values. The X values are in the C column and the Y values are in the D column. These values are divided into three sections. The first section of X values contain the high (1) and low (0) values for the X range. These values match up with the various Y values to plot the log grid lines which are displayed as horizontal lines. The second section has the X values for the vertical grid lines, finally the third section has the X values for the plotted data.

The breaks in the data created by spaces allow the plot to stop in one location and then restart at a new location. To enter these values is tedious but can be done in a few minutes. One could create a macro to perform this task.

Enter the following values into the C and D columns:

C1: +$B$2	D1: +$B$2		
C2: +$B$2	D2: +$B$27		
C4: +$B$3	D4: +$B$2		
C5: +$B$3	D5: +$B$27		
C7: +$B$4	D7: +$B$2		
C8: +$B$4	D8: +$B$27		
C10: +$B$5	D10: +$B$2		
C11: +$B$5	D11: +$B$27		
C13: +$B$6	D13: +$B$2		
C14: +$B$6	D14: +$B$27		
C16: +$B$7	D16: +$B$2		
C17: +$B$7	D17: +$B$27		
C19: +$B$8	D19: +$B$2		
C20: +$B$8	D20: +$B$27		
C22: +$B$9	D22: +$B$2		
C23: +$B$9	D23: +$B$27		
C25: +$B$10	D25: +$B$2		
C26: +$B$10	D26: +$B$27		
C28: +$B$11	D28: +$B$2		
C29: +$B$11	D29: +$B$27		
C31: +$B$12	D31: +$B$2		

```
C32:  +$B$12      D32:  +$B$27
C34:  +$B$13      D34:  +$B$2
C35:  +$B$13      D35:  +$B$27
C37:  +$B$14      D37:  +$B$2
C38:  +$B$14      D38:  +$B$27
C40:  +$B$15      D40:  +$B$2
C41:  +$B$15      D41:  +$B$27
C43:  +$B$16      D43:  +$B$2
C44:  +$B$16      D44:  +$B$27
C46:  +$B$17      D46:  +$B$2
C47:  +$B$17      D47:  +$B$27
C49:  +$B$18      D49:  +$B$2
C50:  +$B$18      D50:  +$B$27
C52:  +$B$19      D52:  +$B$2
C53:  +$B$19      D53:  +$B$27
C55:  +$B$20      D55:  +$B$2
C56:  +$B$20      D56:  +$B$27
C58:  +$B$21      D58:  +$B$2
C59:  +$B$21      D59:  +$B$27
C61:  +$B$22      D61:  +$B$2
C62:  +$B$22      D62:  +$B$27
C64:  +$B$23      D64:  +$B$2
C65:  +$B$23      D65:  +$B$27
C67:  +$B$24      D67:  +$B$2
C68:  +$B$24      D68:  +$B$27
C70:  +$B$25      D70:  +$B$2
C71:  +$B$25      D71:  +$B$27
C73:  +$B$26      D73:  +$B$2
C74:  +$B$26      D74:  +$B$27
C76:  +$B$27      D76:  +$B$2
C77:  +$B$27      D77:  +$B$27
C85:  +$B$2       D85:  +$B$40
C86:  +$B$27      D86:  +$B$40
C88:  +$B$2       D88:  +$B$41
C89:  +$B$27      D89:  +$B$41
C91:  +$B$2       D91:  +$B$42
C92:  +$B$27      D92:  +$B$42
C94:  +$B$2       D94:  +$B$43
C95:  +$B$27      D95:  +$B$43
C97:  +$B$2       D97:  +$B$44
C98:  +$B$27      D98:  +$B$44
C100: +$B$2       D100: +$B$45
C101: +$B$27      D101: +$B$45
C103: +$B$2       D103: +$B$46
C104: +$B$27      D104: +$B$46
C106: +$B$2       D106: +$B$47
C107: +$B$27      D107: +$B$47
C109: +$B$2       D109: +$B$48
C110: +$B$27      D110: +$B$48
C112: +$B$2       D112: +$B$49
```

```
C113: +$B$27      D113: +$B$49
C115: +$B$2       D115: +$B$50
C116: +$B$27      D116: +$B$50
C118: +$B$2       D118: +$B$51
C119: +$B$27      D119: +$B$51
C121: +$B$2       D121: +$B$52
C122: +$B$27      D122: +$B$52
C124: +$B$2       D124: +$B$53
C125: +$B$27      D125: +$B$53
C127: +$B$2       D127: +$B$54
C128: +$B$27      D128: +$B$54
C130: +$B$2       D130: +$B$55
C131: +$B$27      D131: +$B$55
C133: +$B$2       D133: +$B$56
C134: +$B$27      D134: +$B$56
C136: +$B$2       D136: +$B$57
C137: +$B$27      D137: +$B$57
C139: +$B$2       D139: +$B$58
C140: +$B$27      D140: +$B$58
```

The Data Values

The X values for the data points are placed at the bottom on the C column. The observed X-Time values are entered in cells H2..H6. These values must be transformed and normalized to fit on the plot. This is done in cell C142 with the equation:

```
C142: +(H1-$A$2)/($A$27-$A$2)
```

This equation can then be copied to the other cells with the command:

```
/Copy C142  [Enter]  C143..C146  [Enter]
```

The Y-Molecular Weight values for the data points are placed in a separate column so different symbols can be used for them on the plot. The data points to plot are placed in cells I2..I6. These values must also be transformed and normalized. This is done in cell E142 with the equation:

```
E142: +(@LOG(I1)-@LOG($A$40))/(@LOG($A$58)-
@LOG($A$40))
```

This equation is then copied to the other cells with the command:

```
/Copy E142  [Enter]  E143..E146  [Enter]
```

Figure 14.4 Semi log
plot from Lotus 1-2-3.

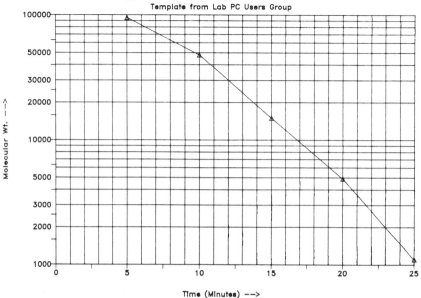

SemiLog Graph Log Mol Wt. vs. Time
Template from Lab PC Users Group

Generating the Plot

We are now ready to generate the plot. Enter the commands:

```
/Graph Type XY X C1..C146 [Enter] A D1..D146
[Enter]
  Options Format A Lines Quit View
```

You should see a plot similar to Figure 14.4. The only things different are that the labels on the axes are not correct and there are no data-labels plotted. This can be corrected by using the Graph Data-labels command and defining a second range to plot.

Completing the Graph

While you are still in the graph command menu, define the second range with the commands:

```
  E E1..E146 [Enter] View
```

We used the E range for no special reason, but we do use a new range so the symbols and lines can be displayed. The data is now plotted on the graph.

Now Quit out of the Graph commands. Enter these values in the F and G columns:

```
F1:  +$A$2      G85:  +$A$40
F16: +$A$7      G88:  +$A$41
F31: +$A$12     G91:  +$A$42
F46: +$A$17     G97:  +$A$44
F61: +$A$22     G112:+$A$49
F76: +$A$27     G115:+$A$50
                G118:+$A$51
                G124:+$A$53
                G139:+$A$58
```

These values will be used as labels on the graph. To use these labels enter the following commands:

```
/Graph Options Scale Y-Scale Hidden Quit
 X-Scale Hidden Quit Quit
```

These commands hide the default axis labels. Now use the Data Labels commands to place new axis labels on the graph using the B and C plot ranges with the commands:

```
B  C1..C146 [Enter]  C  D1..D146  [Enter]
Options Format B Neither C Neither Quit
Data-Labels B F1..F146 [Enter] Below
C G1..G146 [Enter] Left Quit Quit View
```

You should now see the plot shown in Figure 14.4.

General Purpose Template

By setting up the template using absolute references to cells, the template can be used for other applications. For example, by changing the equations in cells B2..B27 and a few other values we can create a log-log plot. Try these changes and you will get the graph shown in Figure 14.5.

First use the Data Fill Command to enter new values in cells A2..A27.

```
/Data Fill A2..A11 [Enter] 1 [Enter] 1
[Enter] 10 [Enter]
/Data Fill A12..A20 [Enter] 20 [Enter] 10
[Enter] 100 [Enter]
/Data Fill A21..A27 [Enter] 200 [Enter] 100
[Enter] 1000 [Enter]
```

Now change the equation in cell B2 to read:

```
B2: +(@LOG(A2)-@LOG($A$2))/(@LOG($A$27)-@LOG($A$2))
```

Figure 14.5 Log-Log plot generated in Lotus 1-2-3.

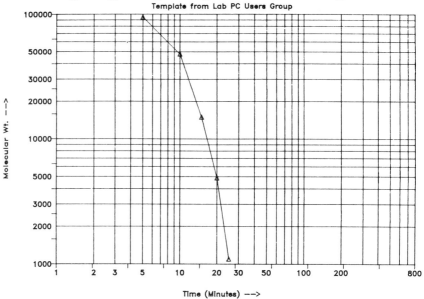

and copy this equation to the other cells in the column with the command:

```
/Copy B2 [Enter] B3..B27 [Enter]
```

Edit the equation in cell C142 to read:

```
C142: +(@LOG(H2)-@LOG($A$2))/(@LOG($A$27)-
@LOG($A$2))
```

and copy this equation to the other data point cells with:

```
/Copy C142 [Enter] C143..C146 [Enter]
```

Erase the previous X-axis labels with the command: /Range Erase F1..F76. New X-axis label values are entered into the following cells:

```
F1:  +$A$2      F4:  +$A$3
F7:  +$A$4      F13  +$A$6
F28: +$A$11     F31: +$A$12
F34: +$A$13     F40: +$A$15
F55: +$A$20     F58: +$A$21
F76: +$A$27
```

Other transformations and plots can be generated using this same template as a starting point.

15

Using Lotus Macros: Combining Instrument Data From Many Files

Combining or importing the data from a large number of individual data files can be automated using a Lotus 1-2-3 macro.

Most laboratory instruments generate a data file for each analytical run they perform. For example, most of the PC based chromatography data systems generate an ASCII (American Standard Code for Information Interchange) file containing the results of an analysis run. These ASCII files can be easily imported into Lotus 1-2-3 by using the /File Import command but the data from only one file is imported for each command. If you want the data from 10 different files you would have to execute the command 10 times.

Lotus Macros to the Rescue

Anytime repetitive keystrokes must be performed to complete a task, a Lotus macro can be written to perform the keystrokes automatically. Lotus macros can also be written which completely automate a task including prompting the user for inputs, printing reports and displaying graphs. Custom menus can also be created in a Lotus macro allowing a novice Lotus or computer user to operate the system by simply making menu selections.

A Lotus macro is a list of keystrokes or Lotus Command Language steps which can be executed by pressing a single keystroke combination. The Lotus Command Language provides macro control commands just like the control commands in a programming language. There are commands, for example, which can call a subroutine and another which can perform "FOR NEXT" or "DO" loops. These commands are contained in "curly bracket" characters when you enter them on a worksheet. The command to perform a FOR NEXT loop is:

```
{FOR counter-location,start,stop,step,loop-loca-
tion}
```

To create a macro, you simply enter the keystrokes or Lotus Commands Language steps as labels into consecutive cells in the worksheet. The first cell in the list of commands is named using the /Range Name Create command. The name given to a macro starts with the symbol \ followed by a letter or number. For example, \A, \H and \8 are typical macro names. To execute the macro, you simply hold down the Alt key and press the letter or number which follows the \ symbol in the macros name.

Automatic Lotus Operations from DOS

A special macro name, \0, is reserved for a macro which will automatically start when the worksheet containing the macro is loaded. A special worksheet name, AUTO123, is reserved for a worksheet which will automatically load whenever Lotus 1-2-3 is run. Thus, by placing a \0 macro in an AUTO123 worksheet, a macro can be automatically started by simply having the computer operator start Lotus 1-2-3 from DOS. These features allow a user to perform any Lotus operations from a DOS Batch file and then return to DOS!

Data Combining Macro

The macro shown in Figure 15.1 can combine the data from a series of files. To use this macro start with a clean worksheet by entering the command:

```
\Worksheet Erase Yes
```

Then type in the macro as shown in Figure 15.1. This macro will work only with Lotus version 2.0 or greater. With a few modifications the macro could work with Lotus version 1. Notice the macro has been placed starting in column AA so that the rest of the worksheet can be used for the data. The labels in the AA column are very important. Be sure you spell the names correctly.

Name the macro and identify other named cells by placing the cursor in cell AA5 and enter the command:

```
/Range Name Labels Right AA5..AA17 [Enter]
```

A second column of labels in cells AG14 through AG19 are similarly turned into named ranges with the command:

	AA	AB	AC	AD	AE	AF	AG	AH
6	\s	{GETNUMBER "Number of Files to Combine: ",PTOT_FILES}~						
7		/rncHERE~{?}~						
8		{home}/fit~						
9		/rncSAVE_DATA~{?}~						
10		~{GOTO}HERE~						
11		/cSAVE_DATA~HERE~						
12		{END}{DOWN}{DOWN}						
13		/rnchere~{BACKSPACE}~						
14		{LET FILE_NUM,0:value}					COUNTER_0	4
15		{FOR COUNTER_0,1,TOT_FILES,1,GETFILE}					PTOT_FILE	4
16		/xq					TOT_FILES	3
17	GETFILE	{HOME}/fit					COUNTER	3
18		{FOR COUNTER,0,FILE_NUM,1,RT_ARROW}					RT_ARROW	{RIGHT}
19		~{GOTO}HERE~					FILE_NUM	3
20		/cSAVE_DATA~HERE~						
21		{END}{DOWN}{DOWN}						
22		/rnchere~{BACKSPACE}~						
23		{LET FILE_NUM,FILE_NUM+1:value}						
24								

Figure 15.1 Macro for combining the data from many files in Lotus 1-2-3. Top part of this figure shows how the macro looks when it is completed on the screen. The bottom part shows the contents of each individual cell in the macro.

```
/Range Name Labels Right AG14..AG19 [Enter]
```

This will assign the labels in the AA and AG columns to the macro commands and open cells in the AB and AH columns. The macro is now ready to use.

Be sure and save your work up to this point with /File Save.

Using the Macro

Before executing the macro you must place all of the files you wish to combine in a single directory. No other files should be in the directory. The macro will blindly read in the files so any files which are present in the directory will be imported. Then assign that directory to be the directory to access with the file command with the keystrokes /File Directory. As written, the macro will only import files with the name extension ".PRN", so make sure your datafiles have this extension.

```
Press: [ALT]-s
```

The macro will prompt you for the number of files to combine as shown in Figure 15.2. Enter the number of files you will be combining. Then the macro will ask you to show where to place the combined data. Press the

Figure 15.2 The Lotus
macro prompts the user
for the number of files to
combine.

```
A1:
Number of Files to Combine: 3

          A         B         C         D         E         F
1
2
3
4
5
6
7
```

Esc key to un-anchor the cursor. Then move the cursor to the position on the worksheet where you want the combined data and press the Enter key. The first file in the directory you have selected is imported and displayed. You then show which cells in the first file to save. The same cells will be saved for each file read. Point to the range of cells you wish to combine as shown in Figure 15.3. If the range has already been selected, you may have to press Esc to unanchor the cursor, move to where you wish to start and again anchor the cursor by pressing the "." key.

Then the macro will read in the number of files you have designated in the first prompt. Each file will have the selected cells copied sequentially below the cell location you selected. Once the last file's data is read and copied, the macro will stop. The data from all of the files you have combined will now be in Lotus 1-2-3 as shown in Figure 15.4.

Retreiving Data

This example macro uses the /File Import Text command to bring straight ASCII text data. You can then use the Data Parse command to parse the text into values. If you have data in ASCII files with values separated by commas and quote marks around text strings you can easily modify the macro to directly import the data. Edit the commands in cells AB8 and AB17 from {HOME}/fit~ to {HOME}/fin~ . Data can also be retrieved from Lotus WK1 files by editing the commands in cells AB8 and AB17 from {HOME}/fit~ to {HOME}/fcce~ (/File Combine Copy Entire-File).

Special Cautions

If you have asked for more files than are present in the directory, the macro will simply loop back to the start and read in the first file again.

```
Enter name: SAVE_DATA                          Enter range: A15..A21

         A          B          C          D          E          F          G          H
3   FILE  1.      METHOD  5.      RUN   2        INDEX    1
4
5   ANALYST:  CRANDELL
6
7   SAMPLE     1      SAMP1
8
9      SA        IS        XF
10  1000.        Ø.        1.
11
12
13  NAME            LB/GAL        RT        AREA BC       RF        RRT
14
15  COMPONENT1         Ø.132     Ø.33     132364 Ø1    1000.       Ø.5
16  COMPONENT2      132358.      Ø.66     132358 Ø1    Ø.001      1.
17  COMPONENT3         123.      Ø.99     132348 Ø1    1.Ø76      1.5
18  COMPONENT4         2Ø2.      1.33     132382 Ø1    Ø.655      2.Ø15
19  COMPONENT5         2Ø3.      1.33     132382 Ø1    Ø.655      2.Ø15
20  COMPONENT6         2Ø4.      1.33     132382 Ø1    Ø.655      2.Ø15
21  COMPONENT7         2Ø5.      1.33     132382 Ø1    Ø.655      2.Ø15
22
```

Figure 15.3 Selecting the range of cells to save from each file. The same range of cells will be stored from each file. A smarter macro could "compute" which cells to store and not just use the same range.

The location where the next set of data will be stored is selected by simply moving to the bottom of the current column of entries and then moving down one more cell. If the wrong column of data is used for this purpose, it is possible that needed data would be overwritten.

Always test a new macro you have written with data you have backed up. You can never predict what may happen with a new macro.

No Frills Macro

This is a real no frills macro. You should look upon it as a starting place for a more specific and useful macro you can create for your specific data application. This macro simply retrieves the data from each file in the directory. Your own custom macro could also include other Lotus commands for performing graphics, data management and data reporting.

	A	B	C	D	E	F	G
30	COMPONENT1		0.132	0.33	132364 01	1000.	0.5
31	COMPONENT2	132358.		0.66	132358 01	0.001	1.
32	COMPONENT3	123.		0.99	132348 01	1.076	1.5
33	COMPONENT4	202.		1.33	132382 01	0.655	2.015
34	COMPONENT5	203.		1.33	132382 01	0.655	2.015
35	COMPONENT6	204.		1.33	132382 01	0.655	2.015
36	COMPONENT7	205.		1.33	132382 01	0.655	2.015
37	COMPONENT1		0.001	0.33	1123 01	1000.	0.5
38	COMPONENT2	1234.		0.66	1234 01	0.001	1.
39	COMPONENT3	11.473		0.99	12345 01	1.076	1.5
40	COMPONENT4	188.482		1.33	123456 01	0.655	2.015
41	COMPONENT5	189.482		1.33	123456 01	0.655	2.015
42	COMPONENT6	180.482		1.33	123456 01	0.655	2.015
43	COMPONENT7	187.482		1.33	123456 01	0.655	2.015
44	COMPONENT1		0.002	0.33	2364 01	1000.	0.5
45	COMPONENT2	4358.		0.66	4358 01	0.001	1.
46	COMPONENT3	11.476		0.99	12348 01	1.076	1.5
47	COMPONENT4	202.109		1.33	132382 01	0.655	2.015
48	COMPONENT5	206.109		1.33	132382 01	0.655	2.015
49	COMPONENT6	208.109		1.33	132382 01	0.655	2.015
50	COMPONENT7	209.109		1.33	132382 01	0.655	2.015
51							

Figure 15.4 Result of combining the files. The data from many files is now combined in Lotus 1-2-3 and ready for further analysis.

This application has shown how Lotus macros provide a way to automate any procedure you perform using Lotus 1-2-3. In other applications in this book, other macros will be developed.

16

Generating Control Charts

Creating control charts of your control sample data is an excellent application you can perform using Lotus 1-2-3. All three major parts of Lotus 1-2-3 can be utilized. Control sample data can be stored in a Lotus database. The necessary calculations can be performed on the Lotus spreadsheet and the charts generated using Lotus graphics.

This application will describe how control charts can be generated. These charts are utilized in many industries to document data quality. This application will describe how these charts are created and maintained in a pharmeceutical company's analytical instrument laboratory. This lab analyzes the raw chemicals, production intermediate compounds and final products and formulations. The raw chemicals and production intermediate compounds are tested for purity. The final products and formulations are tested for purity and shelf life under various environmental conditions including high temperature and high humidity. These tests are performed by using gas and liquid chromatography to separate, identify, and quantitate the compounds.

Control Charting in a Pharmaceutical Analysis Laboratory

Maintaining control charts for all analytical instruments documents data accuracy and should lead to better data quality. Using the techniques in this application you can create and maintain control charts cost-effectively.

In the pharmaceutical industry, accurate, quality data is the cornerstone of all good laboratory practice and decisions. It does not matter if you are following one reaction in a 10-ml flask or several thousand tons

Figure 16.1 Control
chart plotting data from
control samples. The next
new data point is plotted to
see if it is within the
specified range of
acceptable values. Lines
are displayed for the
average, average plus two
standard deviations and
average minus two
standard deviations.

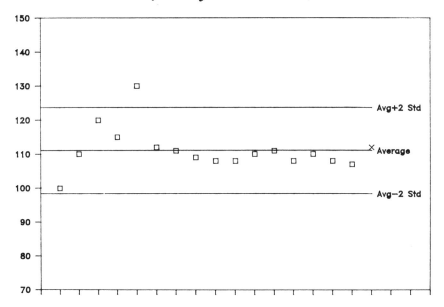

Data Quality Control Chart

of products from a series of reaction vessels, if your data is inaccurate
you are going to make the wrong decisions and have problems. A common method used to track the accuracy of an analysis is to analyze a
control sample whose result should be constant and compare the current
result with historical values. A quick way to insure new data is valid is
to require that the control sample's current value is within predeter-
mined limits. Usually new control sample values must fall within one
or two standard deviations from the average of previous control samples or a reason must be identified for why the results were not consis-
tant.

Control charts, like the one shown in Figure 16.1, are graphs generated
from the analysis of control samples. Usually these charts display a
certain number of historical data points, and then lines showing the
average, and the acceptable control sample ranges. If a new value falls
outside of the specified range, it is easily detected by looking at the
chart.

Government regulatory agencies like the Environmental Protection
Agency and the Food and Drug Adminstration require, under certain
programs, the creation of control charts to validate reported results.
Results obtained not supported with proof of the quality of the data
will not be accepted. Even without these regulatory requirements, any

results obtained in a laboratory should have some level of documented accuracy. Maintaining a control chart on each of your instruments and analysis type can be the first step you can take to document and measure the accuracy of the data you are reporting.

Cost-Effective Application for a PC

Capturing, storing, displaying and plotting control chart data is an excellent, cost-effective application for a personal computer (PC). The cost of a PC can easily be recovered through the savings in labor and accuracy of the process. Most of the requirements for producing control charts by hand are labor intensive. A single chart created by hand can easily take hours to calculate and plot. Because repetitive hand calculations are tedious, there is a high chance of error in performing the calculations or while entering the raw data.

PCs can perform these laborious repetitive tasks without error. PCs also have the ability to communicate control sample data from various instruments to other computers, so data quality can be monitored throughout the laboratory and included in reports. This communication capability saves other people in your organization time since in most cases, the data you generate will eventually be used in a computer system of some type.

The PC is also located directly "where the action is". If there is a situation where poor data is being generated, the problem can be quickly diagnosed and resolved. This will save you from collecting many hours or days of poor data.

Creating a Control Chart

This application will create the control chart shown in Figure 16.1.

First the data to be plotted must be entered into the A column:

A3:100	A7:130	A11:108	A15:108	A19:112
A4:110	A8:112	A12:108	A16:110	
A5:120	A9:111	A13:110	A17:108	
A6:115	A10:109	A14:111	A18:107	

Naming a Range of Cells

Individual cells or ranges of cells can be given a name. The result is a "named-range". Whenever Lotus asks for a cell or range of cells the name you have given the named-range can be entered. Using named-ranges documents what the values in the range represent. Named-ranges are also very valuable when you set up a spreadsheet which refers to a range of cells which may change as the spreadsheet is used.

If named ranges are used, only the cells identified in the named range need to be changed, all of the individual equations using the named range will be automatically updated. If you do not use a named range, you would have to update all of the equations.

Let's create a range name for our original data.

Move the cursor to cell A3.

```
/ Range Name Create DATA [Enter] A3..A18
[Enter]
```

Note we did not include the data point in cell A19 in our range. That is our last control sample point and we want to compare it to our current data. That point will get special treatment and plotted in its own range.

To draw a horizontal line, the same value is placed in a column of numbers. We would like to have the average of our data plotted.

Move the cursor to cell B3.

```
B3: +@AVG($data)
```

Copy the equation in B3 to the cells in the B column.

```
/ Copy B3 [Enter] B3..B19 [Enter]
```

This creates a column of data all with the value of the average of our control sample data. Note that if we need to add a new data point, the original named range "data" need only be changed and all of the cells in column B will also be updated.

Graphing

```
/ Graph A A3..A18 [Enter] B B3..B18 [Enter]
View
```

The graph should look like Figure 16.2. Now we will change the format so the data will be displayed as symbols only and the average line is lines only. We will also change the Y scale.

```
Options Format A Symbols B Lines Quit Quit View
Options Scale Y-Scale Manual Lower 0 [Enter]
Upper 150 [Enter] Quit Quit View
```

the graph will look like Figure 16.3.

```
Quit
```

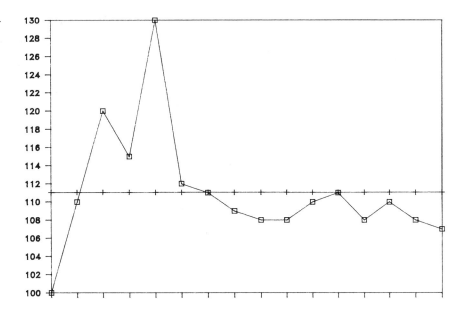

Figure 16.2 First primitive graph in creating a control chart.

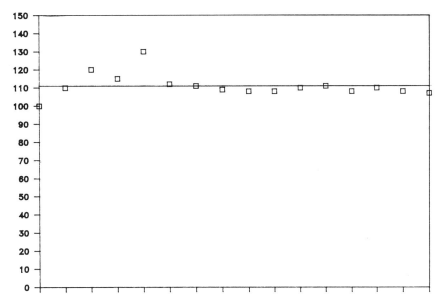

Figure 16.3 Intermediate graph in creating a control chart.

Plotting Blank Cells

To make the graph more eye appealing, we will add some blank cells to our graph ranges for plotting the data. The average line will be extended so it will continue to be plotted across the entire screen.

First copy the equations in the B column to a new location. Then move the "latest control value" to a different column so it will be displayed with a different symbol and not included in the average and standard deviation calculations.

```
/ Copy B3 [Enter] B2..B2 [Enter]
/ Move A19 [Enter] H19 [Enter]
Move to: A2
```

and redefine the graph ranges.

```
/ Graph A [ESC] A2..A19 [Enter] B [ESC]
B2..B19 [Enter]
Quit
```

Adding the Other Lines

To add the control lines, two more columns of data must be created and plotted. Enter the following equations in cells C2 and D2.

```
C2: +@AVG($DATA)-@STD($DATA)
D2: +@AVG($DATA)+@STD($DATA)
```

Now copy these equations to the other cells in the C and D columns.

```
/ Copy C2..D2 [Enter] C3..D19 [Enter]
```

Define two new ranges to plot and rescale the Y axis.

```
/Graph C C2..C19 [Enter] D D2..D19 [Enter]
View Options Format C Lines D Lines
Quit Scale Y-Scale Lower 70 [Enter] Quit Quit
View
```

The graph will look like Figure 16.4.

Changing to Two Standard Deviations

To show how easy it is to change the position of the control lines, let's change them from one standard deviation to two standard deviations. First edit the current cell contents and then copy the new formulas to the other cells in the columns.

Quit out of the Graph commands.

```
Move to: C2
Press: [EDIT] (usually the [F2] key)
```

Move the cursor to the @ sign in front of STD with [LEFT ARROW].

```
Type: 2* [ENTER].
```

Figure 16.4
Intermediate graph in
creating the control chart.

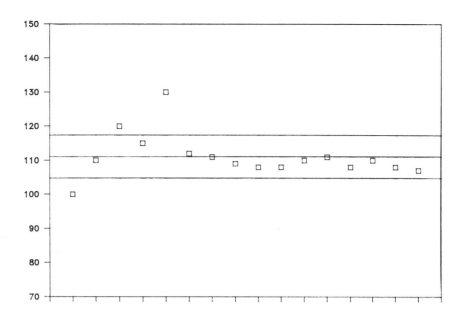

```
Move to: D2
Press: [EDIT]
```

Move the cursor to the @ sign in front of STD with [LEFT ARROW].

```
Type: 2* [ENTER].
```

Copy these edited equations to the rest of the cells in the C and D columns.

```
/ Copy C2..D2 [Enter] C3..D19 [Enter]
```

To graph the new control sample value with a different symbol we define a new range of cells to plot.

```
/ Graph E H2..H19 [Enter] View
```

Now we want to add labels to the end of the lines. First we will add more blank cells to make room for the labels on the plot. Then we will use the Data-Labels command to identify which cell contents to place on the graph.

```
A A2..A22 [Enter] B B2..B22 [Enter] C
C2..C22 [Enter] View
Quit
```

Enter the following labels in the cells.

```
E19: 'Average
F19: 'Avg-2 Std
G19: 'Avg+2 Std
```

Now identify the cells to plot.

```
/ Graph Options Data-Labels B E2..E19 [Enter]
Right
C F2..F19 [Enter] Right D G2..G19 [Enter]
Right
Quit Quit View
```

Add in titles and axis labels to complete the control chart.

The control chart should become a familiar data representation in your laboratory to track data quality. You can easily add new data points to the control charts we have generated.

The next application will describe how Lotus data management commands can be used to store and retrieve control chart data.

17

Maintaining Control Chart Data with Lotus 1-2-3

Maintaining control charts for all of your instruments will document your data accuracy and should lead to better data quality. Using the techniques in this application, a PC and Lotus 1-2-3 you can create and maintain control chart data cost-effectively.

Data Management and LIMS

Data management programs are one part of a Laboratory Information Management System (LIMS). A LIMS has five major components which all must work together to create a successful system. The five components listed by order of importance are: people, procedures, data, programs, and computer hardware. You are already familiar with data, programs, and computer hardware. Let's look at the other components of a LIMS in more detail.

People are the Reason

People are the reason a LIMS is developed. The LIMS is being developed to provide answers for people who must make decisions. Other people will design, create, and operate the LIMS. Still others will enter and check data. People involved with a LIMS seem to fall into three general categories: decision makers, technical users, and clerical users. Many times in a PC based LIMS, a single individual may fall into one, two, or all three categories.

The decision makers are the ones with the questions which need answers. They are the main reason why the LIMS is created. They need the answers from the LIMS so they can make the right decisions. They usually have little interest in computing or technical decisions on how to implement a LIMS. Their main interest is to ensure the LIMS will provide the right answers to their questions. Occasionally, a decision maker will want to ask questions not specifically programmed in the LIMS. For these occasions, the LIMS must be easy to use and flexible in the way questions can be asked.

Technical users are the ones who design and implement the LIMS. They must be sure that all five components are working together properly and the right answers are being generated by the LIMS. Their main focus is on system design, data and program back-up, system documentation, training, and recovery from possible system errors.

Clerical users have the huge task of entering and verifying the data going into the LIMS. Each piece of data entering the LIMS must either be entered through a keyboard by a clerical user or, when possible electronically transfered from an instrument. Even when data is electronically transfered, a clerical user must verify the data is correct and entered in the proper form and time.

Setting Up Procedures

Procedures provide the pipes which allow the data to flow properly into and through the LIMS. If procedures are not established and followed, it will be like leaving a section of pipe out of a plumbing system, a sure disaster. Decision makers must know how and when to ask their questions. Clerical users must have clear instructions on how to operate the computer, enter data properly, and what to do if something goes wrong.

Procedures are designed and documented by the technical users. The technical users must teach the procedures to both decision makers and clerical users. Setting up, documenting, and getting people involved in a LIMS to follow procedures is the biggest single task needed to implement a LIMS. If procedures are not followed, the LIMS will not provide the right answers.

This application will describe how Lotus 1-2-3 data management commands can manage control chart historical data. Even if your laboratory uses other data management programs either on PCs, mini- or mainframe computers, you can still utilize the techniques in this application. Most data management programs like RBase 5000 and dBASE III do not include the ability to generate graphs. By teaming up Lotus 1-2-3 with your data manager you can create an easy to use system for control charting. Lotus 1-2-3 can easily read data from most other PC data management programs and even from other mini- and mainframe computers. Lotus 1-2-3 can also read data captured directly from instruments or other computers that is placed in standard file formats.

```
           A         B         C         D         E         F         G         H
 1   Control Chart Data Base
 2   ==================================
 3
 4   Date      Time      Operator  Instrmnt  Analyte   Raw Val   Result    Comment
 5   30-Jun    8.00  GIO           GC1       butane     67800    100.00
 6   30-Jun    8.00  RBK           LC1       mparabin  234590    200.00
 7   30-Jun    9.00  GSE           GC2       butane     68900    100.00
 8   01-Jul    8.00  GIO           GC1       butane     67700     99.85
 9   01-Jul    8.00  RBK           LC1       mparabin  234580    199.99
10   01-Jul    9.00  GES           GC2       butane     68950    100.07
11   02-Jul    8.00  GIO           GC1       butane     67750     99.93
12   02-Jul    8.00  RBK           LC1       mparabin  234690    200.09
13   02-Jul    9.00  GSE           GC2       butane     69050    100.22
14   03-Jul    8.00  GIO           GC1       butane     67650     99.78
15   03-Jul    8.00  RBK           LC1       mparabin  234700    200.09
16   03-Jul    9.00  GSE           GC2       butane     69100    100.29
17   07-Jul    8.00  GIO           GC1       butane     67590     99.69
18   07-Jul    8.00  RBK           LC1       mparabin  234750    200.14
19   07-Jul    9.00  GSE           GC2       butane     69150    100.36
20   08-Jul    8.00  GIO           GC1       butane     67500     99.56
21   08-Jul    8.00  RBK           LC1       mparabin  234800    200.18
22   08-Jul    9.00  GSE           GC2       butane     69200    100.44
23   09-Jul    8.00  GIO           GC1       butane     67500     99.56
24   09-Jul    8.00  RBK           LC1       mparabin  234850    200.22
25   09-Jul    9.00  GSE           GC2       butane     69250    100.51
26
```

Figure 17.1 Data base of control chart data generated in Lotus 1-2-3.

Maintaining the Control Chart Data

Lotus 1-2-3 data management commands can be used to maintain your control chart data. Each time a control sample is analyzed the results of the analysis can be entered into your Lotus data base. A portation of a typical data base for a chromatographic analysis is shown in Figure 17.1. This data base can easily be modified to track data from any type of instrument in your laboratory. If your instrument can communicate with a PC, or has a PC based data system, this data could be captured and placed into Lotus without having to re-key in the data. Techniques for capturing data were described in chapter two of this book.

What is a Data Base?

A data base is a structured collection of information. A telephone book is a common example of a printed data base with thousands of entries, Figure 17.2. Each entry in the telephone book usually has three pieces of information: a name, an address and a telephone number. In data base

Figure 17.2 The telephone book is a printed data base with three major fields per record: a name, an address and a telephone number.

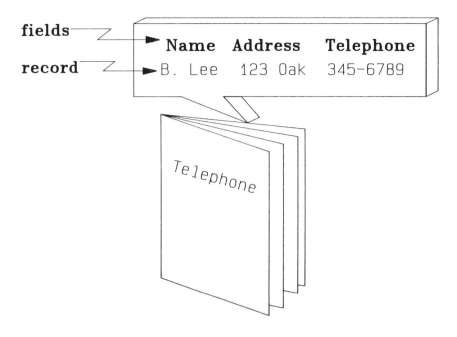

Figure 17.2 The telephone book is a printed data base with three major fields per record: a name, an address and a telephone number.

management terminology, the entries are called records. The pieces of information contained in each record are called fields. Thus, a telephone book has thousands of records, each with information in one of three fields: a name, an address and a telephone number.

We are all familiar with printed data bases and use them successfully in our daily lives. Most are easy to use. Here are some examples of other printed data bases we use often with their normal fields in parenthesis: the card catalog at the library (book title, author, subject), restaurant menu (food item, price), store merchandise catalog (item, description, price), and Merck Index (compound name, formula, description, other physical, pharmacological and chemical data).

Computerized data bases are no more difficult to use than are printed data bases. Computerized data bases have advantages over printed data bases since the access to the data is not limited to a static method of sorting and presenting the data. For example, a telephone book presents its data sorted by the names in alphabetical order. If you know a person's name, you can find their telephone number (as long as it is listed and you are looking in the right telephone book!). But a telephone book is pretty useless if you only have a telephone number and need to know who's number it is. A computerized data base could easily solve this problem by sorting the telephone numbers in numerical order or by making a query to the data base for a direct match. Creating a computerized data base thus provides you with the ability to access your data using many different criteria and sorting capabilities.

Creating a Control Chart Data Base

Our control chart data base contains the data from the analysis of a number of different control samples over a two week period. The results are from three different instruments, gas chromatograph one (GC1), gas chromatograph two (GC2) and liquid chromatograph one (LC1) for two different compounds, butane and mparabin. In an actual application you may want to create separate data bases for each instrument or even each compound on each instrument. In the data base shown in Figure 17.1, we have 8 fields:

```
DATE - Date the control sample was analyzed
TIME - Time the control sample was analyzed
OPERATOR - Operator running the control sample
INSTRMNT - Instrument used to run the control
sample
ANALYTE - Compound analyzed
RAW VAL - Raw instrument value
RESULT - Result reported after considering
calibration data
COMMENTS - Any comments about the analysis
```

Other fields you may consider adding to your own control chart data base could be:

```
EXPECTED RESULT - Value expected from the analysis
CALIBRATION DATA - Calibration data or calibration
curve coefficients.
```

Data Management with Lotus 1-2-3

To perform data management in Lotus 1-2-3 each field is a spreadsheet column. Each of the data base records is a spreadsheet row. When we analyze a control sample, the data is logged into the data base by filling in values for each field across a row. Data is easily entered using the normal data entry and editing methods available in Lotus 1-2-3.

To set up a data base you simply enter field names at the top of a column and then enter your data in the rows below. To create this spreadsheet application yourself all you need is a copy of Lotus 1-2-3 Version 1A or Release 2.0. To create the data base in Figure 17.1 do the following.

Start with a blank spreadsheet. To insure the spreadsheet is initialized perform a Worksheet Erase with the command:

```
/Worksheet Erase Yes
```

Starting at cell A1 enter:

```
A1: 'Control Chart Data Base
Move to: A2
Type: \=
```

Copy the equal signs to cells B2 and C2 with the command:

```
/Copy A2 [Enter] B2..C2 [Enter]
```

Enter the Field headings, in row 4, in the following cells:

```
A4: 'Date        B4: 'Time        C4: 'Operator
D4: 'Instrmnt
E4: 'Analyte     F4: 'Raw Val     G4: 'Result
H4: 'Comments
```

Now we set data display formats and column widths. For this application the A column uses a Date format while columns B and G use a fixed two decimal place format. Lotus allows many other display formats including scientific notation, percentage and currency. Column widths were left at the default 9 characters. Column widths can be from one to 72 characters. The formats shown are set using the Range command:

```
/Range Format A5..A200 [Enter] Date DD-MMM
  Format B5..B200 [Enter] Fixed 2 [Enter]
/Range Format G5..G200 [Enter] Fixed 2 [Enter]
```

The data is then entered as shown in Figure 17.1. The dates in column A are entered using the @DATE(YY,MM,DD) function. The YY is the year, MM the month and DD the day of the month. The date function for July 4, 1986 is @DATE(86,7,4). The @DATE function enters a number into the cell which is the number of days since 31-Dec-1899. When you use a DATE display format, you can display the date in one of three ways, Day-Month-Year (04-Jul-86), Day-Month (04-Jul) or Month-Year (Jul-86). The DATE value can then be tested numerically using the @DATE function (see Using Numerical and Date Functions in this chapter).

Extracting Data from the Data Base

To extract data out of the data base to be plotted in control charts requires using a few Lotus data management commands. First, we identify three special ranges of cells on the spreadsheet. These ranges are called the INPUT, OUTPUT and CRITERION ranges. The INPUT range is the range of cells containing the data in the data base. The OUTPUT range is the range of cells where the results of the data extraction will be placed. The CRITERION range is where the types of data to be extracted are described.

For this example the INPUT range is A4..H200, the OUTPUT range is I12..P208 and the first CRITERION range is I4..P5.

We set up these ranges by doing the following. The first row in each of these ranges is a row of field name labels. This row of field labels is required as the first row in each of the three ranges. Since we have already entered the field labels in row 4 we can simply copy the labels to the other locations. Enter:

```
/Copy  A4..H4  [Enter]  I4..P4   [Enter]
/Copy  A4..H4  [Enter]  I12..P12 [Enter]
```

Now identify the various ranges with the commands:

```
/Data Query Input A4..H200 [Enter] Criteria
I4..P5 [Enter]
 Output I12..P208 [Enter] Quit
```

Now we add a few labels to dress up the screen and direct our attention to the correct parts of the spreadsheet we have created.

In cell

```
I1:  Query
I2:  \=
I10: Output
I11: \=
```

Query the Data Base

Now we can extract the control chart data from our data base by using the Data Query command. To extract all of the data for butane we enter the label butane under the ANALYTE field in the CRITERION range in cell M5. The row just below the field labels is called the Query Line. We now enter the command:

```
/Data Query Extract
```

All records in the data base which contain butane in the ANALYTE field are now extracted from the data base and placed into the OUTPUT range as shown in Figure 17.3. After you have entered the query command once, you can perform another query by simply pressing the [Query] key, which is usually the [F7] key.

```
         I         J         K         L         M         N         O         P
 1   Query
 2   =========
 3
 4   Date      Time      Operator  Instrmnt  Analyte   Raw Val   Result    Comment
 5                                           butane
 6
 7
 8
 9
10   Output
11   =========
12   Date      Time      Operator  Instrmnt  Analyte   Raw Val   Result    Comment
13   30-Jun    8.00 GIO            GC1       butane    67800     100.00
14   30-Jun    9.00 GSE            GC2       butane    68900     100.00
15   01-Jul    8.00 GIO            GC1       butane    67700      99.85
16   01-Jul    9.00 GES            GC2       butane    68950     100.07
17   02-Jul    8.00 GIO            GC1       butane    67750      99.93
18   02-Jul    9.00 GSE            GC2       butane    69050     100.22
19   03-Jul    8.00 GIO            GC1       butane    67650      99.78
20   03-Jul    9.00 GSE            GC2       butane    69100     100.29
21   07-Jul    8.00 GIO            GC1       butane    67590      99.69
22   07-Jul    9.00 GSE            GC2       butane    69150     100.36
23   08-Jul    8.00 GIO            GC1       butane    67500      99.56
24   08-Jul    9.00 GSE            GC2       butane    69200     100.44
25   09-Jul    8.00 GIO            GC1       butane    67500      99.56
26   09-Jul    9.00 GSE            GC2       butane    69250     100.51
27
```

Figure 17.3 Extracting control chart data to be plotted from the data base using a compound name as a single criteria.

Playing a Text Wildcard

Our first query used an exact match of text or labels. The labels must match exactly including upper and lower case characters. Queries can contain "wildcard" symbols. The * symbol is a wildcard for multiple symbols while the ? symbol is a wildcard for a single symbol. Like in a card game, a wildcard symbol can replace any character or multiple characters when trying to make a match. For example using the data base displayed in Figure 17.1, if you used the label G* in cell K5 and executed a query, you would obtain in the output range the entries with the Operators GIO and GSE. The G* requests records which contain for the Operator field operator initials starting with the letter capital G and any other letters can follow. The ? can replace any single letter.

Thus if you place the label GC? in cell L5, you will extract records from all of the results with GC and then any character like GC1 and GC2. Be sure when you try these queries all other cells on the query line are empty.

Using More Criteria in the Query

Normally you will want to view and plot various combinations of data. You may want to view the data for a specific instrument and compound over a specific time period or a specific instrument, compound, and operator over a specific time. Extracting this data can be done by adding additional criteria to the Query Line. If you have a single Query Line, each of the criteria on the line must be present before a record is extracted. In data management terms the query criteria are ANDed. Thus, if we place the label GC1 in cell L5 in addition to the label butane in M5, only the results which have the label butane for ANALYTE AND GC1 for INSTRMNT will be placed in the OUTPUT range as shown in Figure 17.4.

	I	J	K	L	M	N	O	P
1	Query							
2	════							
3								
4	Date	Time	Operator	Instrmnt	Analyte	Raw Val	Result	Comment
5				GC1	butane			
6								
7								
8								
9								
10	Output							
11	════							
12	Date	Time	Operator	Instrmnt	Analyte	Raw Val	Result	Comment
13	30-Jun	8.00	GIO	GC1	butane	67800	100.00	
14	01-Jul	8.00	GIO	GC1	butane	67700	99.85	
15	02-Jul	8.00	GIO	GC1	butane	67750	99.93	
16	03-Jul	8.00	GIO	GC1	butane	67650	99.78	
17	07-Jul	8.00	GIO	GC1	butane	67590	99.69	
18	08-Jul	8.00	GIO	GC1	butane	67500	99.56	
19	09-Jul	8.00	GIO	GC1	butane	67500	99.56	
20								

Figure 17.4 Data base query using multiple criteria. Both the ANALYTE entry AND the INSTRMNT entry in a record must match for the data to be extracted from the data base and placed in the OUTPUT range.

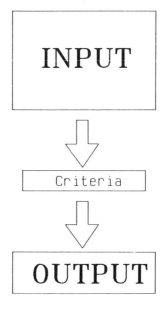

Using Numerical and Date Intervals

To extract the data from a specific time period or over a specific range of numerical values, you use other criteria statements. For these criteria statements you use the cell location just below the field label in equations you construct to describe the criteria you wish to use. For example, to extract only results entered after July 3, you enter in cell I5 :

```
+A5>@DATE(86,7,3)
```

The cell identifier A5 is used in the equation because it is the cell location just below the Date field label (or the first cell containing data). All arthmetic and logical operators can be included in these equations to precisely identify the data you wish to extract. The common operators are shown in Table 1.

With these functions you can construct the numerical queries you will need. For example to extract the data collected from June 30, 1986 til July 3, 1986 you enter the following equation into cell I5:

```
+A5>=@DATE(86,6,30)#AND#+A5<=@DATE(86,7,3)
```

Extracting Data Using a Logical OR

Data can also be extracted from our data base using a logical OR. The data extraction operation is like performing a "filtration" of the entire data base, Figure 17.5. The Query Line is our filter. If the contents of a data base record does not match each criteria in the Query Line filter, the record falls through and is not extracted from the data base nor displayed in the output range. We are not limited to a single Query Line or filter. You can include many more filters by simply defining the CRITERION range to have additional Query Lines. Let us say we want to extract a data record whether it is from GC1 OR GC2. We first expand the CRITERION range to include an additional line.

/Data Query Criterion **I4..P6** **[Enter]**

Now we place values into the two Query Lines as shown in Figure 17.6. Be cautious when using two or more Query Lines because a blank query line will extract every record of a data base.

	I	J	K	L	M	N	O	P
1	Query							
2	=======							
3								
4	Date	Time	Operator	Instrmnt	Analyte	Raw Val	Result	Comment
5				GC1	butane			
6				GC2				
7								
8								
9								
10	Output							
11	=======							
12	Date	Time	Operator	Instrmnt	Analyte	Raw Val	Result	Comment
13	30–Jun	8.00	GIO	GC1	butane	67800	100.00	
14	30–Jun	9.00	GSE	GC2	butane	68900	100.00	
15	01–Jul	8.00	GIO	GC1	butane	67700	99.85	
16	01–Jul	9.00	GES	GC2	butane	68950	100.07	
17	02–Jul	8.00	GIO	GC1	butane	67750	99.93	
18	02–Jul	9.00	GSE	GC2	butane	69050	100.22	
19	03–Jul	8.00	GIO	GC1	butane	67650	99.78	
20	03–Jul	9.00	GSE	GC2	butane	69100	100.29	
21	07–Jul	8.00	GIO	GC1	butane	67590	99.69	
22	07–Jul	9.00	GSE	GC2	butane	69150	100.36	
23	08–Jul	8.00	GIO	GC1	butane	67500	99.56	
24	08–Jul	9.00	GSE	GC2	butane	69200	100.44	
25	09–Jul	8.00	GIO	GC1	butane	67500	99.56	
26	09–Jul	9.00	GSE	GC2	butane	69250	100.51	
27								

Figure 17.6 Query using a logical OR. Two query lines are defined. If a record matches either the first query line OR the second query line, it is extracted and placed in the OUTPUT range.

As you can see, your control chart data can be managed using the Lotus 1-2-3.

Many Options

This application has shown how control chart data can be managed using Lotus 1-2-3. Many other graphing options can be used to create different charts and graphs from the data. Other data management techniques can be used to streamline the application.

18

1-2-3 Laboratory Analysis Logbook

Modern scientific instruments and product test equipment have the capability to send their results or raw data to an IBM PC. This application describes how the data can be managed and displayed using the data management and graphics capabilities of Lotus 1-2-3. An analysis logbook is constructed for data management.

The logbook you construct can be used to track the results of tests performed on any final or intermediate manufactured product. It can be easily modified to perform a number of functions. Data is entered into the logbook in two ways, manually through the keyboard and by "importing" data files captured directly from the instruments. This data can then be analyzed, correlated and summarized using 1-2-3's data management and graphics commands.

Building the Analysis Logbook

Let's set up our logbook for the analysis results. The logbook has two parts, data log and query. The data log is a list of the sample analysis information including a reference number, date the sample was received, who submitted the sample, type of analysis, analyte or feature measured in the sample, analysis raw value, computed result, date analyzed and a memo for other information. The data query section is for summarizing and extracting data from the data log information.

The explanation for setting up the logbook will attempt to use the same terminology used by the 1-2-3 instruction manual for the various commands.

The first step is to build the Data Log Input form. Make sure you start
with a clean worksheet by giving the command:

```
/Worksheet Erase Yes
```

Enter these labels in the following cells:

```
A1: '*** Laboratory Analysis Logbook ***
A5: 'Data Log
A6: \=
```

Copy the contents of A6 to cells B6 through I6 with the command:

```
/Copy A6 [Enter] B6..I6 [Enter]
```

Now enter the field names in row 7.

```
A7: "Ref no.
B7: "Rec Date
C7: 'Submitted by
D7: 'Analysis
E7: 'Analyte
F7: 'Raw Value
G7: 'Amount
H7: 'Date Anal
I7: 'Memo
A8: \-
```

Use the Copy command to complete the line from B8 through I8.

```
/Copy A8 [Enter] B8..I8 [Enter]
```

Now reset the column width of column C to 15 characters. Got to any cell
in column C and give the command:

```
/Worksheet Column-Set 15 [Enter]
```

Using the same command set the column widths of columns D and E to a
width of 8 characters each. Set columns F and G to 12 characters, column
H to 9 characters and column I to 15 characters.

Most of the numbers on the worksheet will be displayed with two decimal places. Set the entire worksheet for this format with the command:

```
/Worksheet Global Format Fixed 2 [Enter]
```

```
         A       B         C          D     E      F        G        H         I
1    ††† Laboratory Notebook
2
3
4
5    Data Log
6    ==========================================================================================
7    Ref no. Rec Date Submitted by      Analysis Analyte Raw Value    Amount    Date Anal    Memo
```

Figure 18.1 Data Log Input Section

The "Ref no." and "Raw Value" columns should display their data as integers. Reformat columns A and F as fixed with no decimal places with the command:

/Range Format Fixed 0 [Enter] A9..A500 [Enter]

Your worksheet should look like Figure 18.1 and you are ready to enter some data. First, we should save our work with the command:

/File Save PCLAB [Enter]

To show how the logbook can be used we'll enter some data from a typical chemical analysis laboratory. This lab performs six types of analyses:

```
GCHROM gas chromatography
LCHROM liquid chromatography
ICP Inductively Coupled Plasma Spectroscopy
UV Ultra-violet Spectroscopy
IR Infrared Spectroscopy
GPC Gel Permeation Chromatography
```

on seven different compounds or elements: propane, asprin, Mercury (Hg), butane, polyv (polyethylene), methane and ethane. When a sample is received it is entered into the data log by filling in a reference number, date received, the submitter's name analysis type and analyte. When the analysis is completed, the raw value, amount and date analyzed are filled in along with any comments in the memo section. When calibration runs are performed on an instrument, the results are also logged. The submitter's name is replaced with Calibration for these entries.

```
         A        B          C              D          E          F          G            H            I
 1  *** Laboratory Notebook
 2
 3
 4
 5  Data Log
 6  ==================================================================================================
 7  Ref no. Rec Date Submitted by       Analysis Analyte Raw Value  Amount      Date Anal    Memo
 8      101 8308.01 Smith               GCHROM   propane   201045    400.00    8308.04
 9      102 8308.01 Jones               LCHROM   asprin     45346    385.00    8308.04
10      103 8308.02 Green               ICP      Hg       2314560     25.60    8308.05
11      104 8308.03 Brown               ICP      Hg       2150780     25.50    8308.05
12      105 8308.03 Johnson             UV       asprin     35245    340.00    8308.07
13      106 8308.04 Edger               IR       asprin     40563    358.00    8308.07
14      107 8308.04 Johnson             GCHROM   butane     23568    229.86    8308.08
15      108 8308.04 Weaver              GPC      polyv     342789    469.23    8308.08
16      109 8308.05 Mills               ICP      Hg       1894670     25.30    8308.08
17      110 8309.06 Peterson            GPC      vinyl    3456790    456.00    8308.10 New Entry
18     2001 8307.05 Calibration         GCHROM   propane   230145    500.00    8307.05
19     2002 8307.05 Calibration         GCHROM   methane   150032    500.00    8307.05
20     2003 8307.05 Calibration         GCHROM   ethane    120132    500.00    8307.05
21     2004 8307.12 Calibration         GCHROM   propane   234024    500.00    8307.12
22     2005 8307.12 Calibration         GCHROM   methane   160054    500.00    8307.12
23     2006 8307.12 Calibration         GCHROM   ethane    124098    500.00    8307.12
24     2007 8307.19 Calibration         GCHROM   propane   228023    500.00    8307.19
25     2008 8307.19 Calibration         GCHROM   methane   145045    500.00    8307.19
26     2009 8307.19 Calibration         GCHROM   ethane    118043    500.00    8307.19
27     2010 8307.26 Calibration         GCHROM   propane   232076    500.00    8307.26
28     2011 8307.26 Calibration         GCHROM   methane   155076    500.00    8307.26
29     2012 8307.26 Calibration         GCHROM   ethane    112043    500.00    8307.26
30     2013 8308.02 Calibration         GCHROM   propane   227032    500.00    8308.02
31     2014 8308.02 Calibration         GCHROM   methane   130102    500.00    8308.02
32     2015 8308.02 Calibration         GCHROM   ethane    107021    500.00    8308.02
33     2016 8308.09 Calibration         GCHROM   propane   225021    500.00    8308.09
34     2017 8308.09 Calibration         GCHROM   methane   120034    500.00    8308.09
35     2018 8308.09 Calibration         GCHROM   ethane    100103    500.00    8308.09
36     2019 8308.16 Calibration         GCHROM   propane   220012    500.00    8308.16
37     2020 8308.16 Calibration         GCHROM   methane   125013    500.00    8308.16
38     2021 8308.16 Calibration         GCHROM   ethane     97035    500.00    8308.16
39     2022 8308.23 Calibration         GCHROM   propane   215078    500.00    8308.23
40     2023 8308.23 Calibration         GCHROM   methane   110021    500.00    8308.23
41     2024 8308.23 Calibration         GCHROM   ethane     94002    500.00    8308.23
42     2025 8308.30 Calibration         GCHROM   propane   210037    500.00    8308.30
43     2026 8308.30 Calibration         GCHROM   methane    90103    500.00    8308.30
44     2027 8308.30 Calibration         GCHROM   ethane     90106    500.00    8308.30
45     2028 8309.06 Calibration         GCHROM   propane   208016    500.00    8309.06
46     2029 8309.06 Calibration         GCHROM   methane    74091    500.00    8309.06
47     2030 8309.06 Calibration         GCHROM   ethane     82001    500.00    8309.06
```

Figure 18.2 Analysis Data entered in the Data Log

All dates are entered using the @DATE(YY,MM,DD) function with the first two digits representing the year, the next two the month and the final two characters the day.

Enter the data as shown in Figure 18.2. At this point you should resave the worksheet with the command:

```
/File Save PCLAB [Enter] Replace
```

Names, Names on the Range

The next steps will introduce you to some unique features of 1-2-3. To utilize our data log with the 1-2-3 Data Query and Sort commands we must assign names to various ranges of cells on the worksheet. First, give the entire data log range the name INPUT by going to cell A7 and giving the command:

```
/Range Name Create INPUT [Enter] A7..I500
[Enter]
```

This command names the range of cells from A7 to I500 as INPUT and sets the lower boundary of our data log at row 500. Note the range started on row 7 to include the labels given at the top of each field. These labels must be included in the input range for the Data Query and Sort functions to work correctly.

Now let's set up the Data Query Section. Goto cell J1 and enter "QUERY. Then goto J2 and place a repeating label of equal signs by entering \=. Copy the contents of J2 to K2 through R2 with the command:

```
/Copy J2 [Enter] K2..R2 [Enter]
```

Copy the labels from row 7 of the data log to cells J3 through R3 with the command:

```
/Copy A7..I7 [Enter] J3 [Enter]
```

Now make a copy of cells J2 through R3 in cells J6 through R7 with the command:

```
/Copy J2..R3 [Enter] J6..R7 [Enter]
```

Reset the column widths to match those in the data log section. The Query Section should look like Figure 18.3.

```
       J        K        L         M       N       O        P       Q         R
1
2 Query
3 ================================================================================
4 Ref no. Rec Date Submitted by        Analysis Analyte Raw Value  Amount  Date Anal    Memo
5
```

Figure 18.3 Query Section

To use the Data Query and Sort Commands we must create a few more named ranges. First set up the criterion or selection range by naming cells J3 to R4 as CRITERION with the command:

```
/Range Name Create CRITERION [Enter] J3..R4
[Enter]
```

Then name cells J7 to R500 as OUTPUT with the command:

```
/Range Name Create OUTPUT [Enter] J7..R500
[Enter]
```

Finally name cells J8 to R500 as SORT with the command:

```
/Range Name Create SORT [Enter] J9..R500
[Enter]
```

Notice the CRITERION and OUTPUT ranges included a row of labels but the SORT range does not.

Now you can connect all the named ranges together so they can be used with 1-2-3's Data Query and Sort functions. You must define an input, an output, a sort, and a criterion range. Define these by giving the commands:

```
/Data Query Input INPUT [Enter]
```

Assigns the A7 to I500 range as the input range. Continue in the Data submenu with:

```
Criterion CRITERION [Enter]
```

Assigns the range J3 to R4 as the criterion range.

```
Output OUTPUT [Enter]
```

Assigns the range J7 to R500 as the output range. [Esc] Back to the main Data menu.

```
Sort Data-Range SORT [Enter]
```

Assigns the range J8 to R500 as the sort range.

Finding and Extracting Data

The Data Query Command allows you two ways to select items from your data log. The simplest is the Find method. Suppose you want to view each of the results using ICP analysis.

```
Move to: M4
Type: ICP
```

Now give the command:

```
/Data Query Find
```

The screen will switch to the data log area (INPUT range) with the third entry highlighted. This is the first record in the input range with a 'match' with the ICP criterion. Press [Down] The highlight bar jumps to the next match which is Brown's ICP submitted sample. Pressing [Down] again jumps down to Mills' ICP sample and pressing [UP] puts you back to Green's ICP Sample. When you are finished looking at entries, press [Esc] and you will return to the Query menu.

The other method of Data Query is Extraction. Matching data is displayed in the data output range. First, let's extract out all of the ICP analysis data. Enter the command:

```
/Data Query Extract
```

The matching data entries are displayed as shown in Figure 18.4. Let's extract all of the calibration data.

```
Move to: M4
```

Erase the contents of this cell with the command:

```
/Range Erase M4 [Enter]
```

```
Move to: L4
Type: CALIBRATION
```

```
         J     K      L         M      N      O      P      Q       R
   1
   2 Query
   3 =================================================================
   4 Ref no. Rec Date Submitted by     Analysis Analyte Raw Value Amount Date Anal   Memo
   5                                   ICP
   6 =================================================================
   7 Ref no. Rec Date Submitted by     Analysis Analyte Raw Value Amount Date Anal   Memo
   8    103  8308.02 Green             ICP      Hg      2314560   25.60  8308.05
   9    104  8308.03 Brown             ICP      Hg      2150780   25.50  8308.05
  10    109  8308.05 Mills             ICP      Hg      1894670   25.30  8308.08
  11
```

Figure 18.4 Data Query Extract Results

Then enter the command:

        /Data Query Extract

In a few seconds all of the entries with Calibration will appear in the output range. To be more selective we could have entered values in other labelled fields such as methane in the Analyte column and only those entries with both Calibration and methane would be extracted. Numeric values and formulas can also be used for the match criterion. For example entering the formula +Raw Value>2000000 would yield only three entries, each of the ICP runs since they are the only entries with Raw Values greater than 2000000.

Experiment with the Query capabilities of 1-2-3. Note that once you have set up your matching criterion you can simply press the Query Key which is function key F7 to obtain the queried data.

Sorting and Graphing

Extracted data can also be sorted and graphed. With the Calibration data in the output range enter these sort commands:

        Move to: N8

Enter the commands:

        /Data Sort Primary-Key N8 [Enter] Descending

This assigns the Analyte column as the primary-key.

Secondary-Key **K8** **[Enter]**

This assigns the Received Date as the secondary-key. We have previously defined the sort range so now all we need to do is enter:

G○

7	Ref no.	Rec Date	Submitted by	Analysis	Analyte	Raw Value	Amount	Date Anal	Memo
8	2001	8307.05	Calibration	GCHROM	propane	230145	500.00	8307.05	
9	2002	8307.05	Calibration	GCHROM	methane	150032	500.00	8307.05	
10	2003	8307.05	Calibration	GCHROM	ethane	120132	500.00	8307.05	
11	2004	8307.12	Calibration	GCHROM	propane	234024	500.00	8307.12	
12	2005	8307.12	Calibration	GCHROM	methane	160054	500.00	8307.12	
13	2006	8307.12	Calibration	GCHROM	ethane	124098	500.00	8307.12	
14	2007	8307.19	Calibration	GCHROM	propane	228023	500.00	8307.19	
15	2008	8307.19	Calibration	GCHROM	methane	145045	500.00	8307.19	
16	2009	8307.19	Calibration	GCHROM	ethane	118043	500.00	8307.19	
17	2010	8307.26	Calibration	GCHROM	propane	232076	500.00	8307.26	
18	2011	8307.26	Calibration	GCHROM	methane	155076	500.00	8307.26	
19	2012	8307.26	Calibration	GCHROM	ethane	112043	500.00	8307.26	
20	2013	8308.02	Calibration	GCHROM	propane	227032	500.00	8308.02	
21	2014	8308.02	Calibration	GCHROM	methane	130102	500.00	8308.02	
22	2015	8308.02	Calibration	GCHROM	ethane	107021	500.00	8308.02	
23	2016	8308.09	Calibration	GCHROM	propane	225021	500.00	8308.09	
24	2017	8308.09	Calibration	GCHROM	methane	120034	500.00	8308.09	
25	2018	8308.09	Calibration	GCHROM	ethane	100103	500.00	8308.09	
26	2019	8308.16	Calibration	GCHROM	propane	220012	500.00	8308.16	
27	2020	8308.16	Calibration	GCHROM	methane	125013	500.00	8308.16	
28	2021	8308.16	Calibration	GCHROM	ethane	97035	500.00	8308.16	
29	2022	8308.23	Calibration	GCHROM	propane	215078	500.00	8308.23	
30	2023	8308.23	Calibration	GCHROM	methane	110021	500.00	8308.23	
31	2024	8308.23	Calibration	GCHROM	ethane	94002	500.00	8308.23	
32	2025	8308.30	Calibration	GCHROM	propane	210037	500.00	8308.30	
33	2026	8308.30	Calibration	GCHROM	methane	90103	500.00	8308.30	
34	2027	8308.30	Calibration	GCHROM	ethane	90106	500.00	8308.30	
35	2028	8309.06	Calibration	GCHROM	propane	208016	500.00	8309.06	
36	2029	8309.06	Calibration	GCHROM	methane	74091	500.00	8309.06	
37	2030	8309.06	Calibration	GCHROM	ethane	82001	500.00	8309.06	

Figure 18.5 Calibration Data before Sort

Before the sort the Calibration data is ordered as shown in Figure 18.5. The entire list of thirty entries will be sorted and grouped by Analyte with each set of Analytes ordered by the date they were analyzed as shown in Figure 18.6.

Now this grouped data can be graphed. Escape back to the main menu by entering [Esc]. Select the data to plot:

/Graph **A 08..O17 [Enter]**

7	Ref no.	Rec Date	Submitted by	Analysis	Analyte	Raw Value	Amount	Date Anal	Memo
8	2001	8307.05	Calibration	GCHROM	propane	230145	500.00	8307.05	
9	2004	8307.12	Calibration	GCHROM	propane	234024	500.00	8307.12	
10	2007	8307.19	Calibration	GCHROM	propane	228023	500.00	8307.19	
11	2010	8307.26	Calibration	GCHROM	propane	232076	500.00	8307.26	
12	2013	8308.02	Calibration	GCHROM	propane	227032	500.00	8308.02	
13	2016	8308.09	Calibration	GCHROM	propane	225021	500.00	8308.09	
14	2019	8308.16	Calibration	GCHROM	propane	220012	500.00	8308.16	
15	2022	8308.23	Calibration	GCHROM	propane	215078	500.00	8308.23	
16	2025	8308.30	Calibration	GCHROM	propane	210037	500.00	8308.30	
17	2028	8309.06	Calibration	GCHROM	propane	208016	500.00	8309.06	
18	2002	8307.05	Calibration	GCHROM	methane	150032	500.00	8307.05	
19	2005	8307.12	Calibration	GCHROM	methane	160054	500.00	8307.12	
20	2008	8307.19	Calibration	GCHROM	methane	145045	500.00	8307.19	
21	2011	8307.26	Calibration	GCHROM	methane	155076	500.00	8307.26	
22	2014	8308.02	Calibration	GCHROM	methane	130102	500.00	8308.02	
23	2017	8308.09	Calibration	GCHROM	methane	120034	500.00	8308.09	
24	2020	8308.16	Calibration	GCHROM	methane	125013	500.00	8308.16	
25	2023	8308.23	Calibration	GCHROM	methane	110021	500.00	8308.23	
26	2026	8308.30	Calibration	GCHROM	methane	90103	500.00	8308.30	
27	2029	8309.06	Calibration	GCHROM	methane	74091	500.00	8309.06	
28	2003	8307.05	Calibration	GCHROM	ethane	120132	500.00	8307.05	
29	2006	8307.12	Calibration	GCHROM	ethane	124098	500.00	8307.12	
30	2009	8307.19	Calibration	GCHROM	ethane	118043	500.00	8307.19	
31	2012	8307.26	Calibration	GCHROM	ethane	112043	500.00	8307.26	
32	2015	8308.02	Calibration	GCHROM	ethane	107021	500.00	8308.02	
33	2018	8308.09	Calibration	GCHROM	ethane	100103	500.00	8308.09	
34	2021	8308.16	Calibration	GCHROM	ethane	97035	500.00	8308.16	
35	2024	8308.23	Calibration	GCHROM	ethane	94002	500.00	8308.23	
36	2027	8308.30	Calibration	GCHROM	ethane	90106	500.00	8308.30	
37	2030	8309.06	Calibration	GCHROM	ethane	82001	500.00	8309.06	

Figure 18.6 Calibration Data after Sort

Figure 18.7 Bar Graph of Calibration Data on three compounds

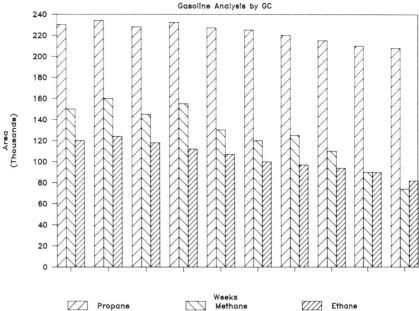

This selects the A range for the propane data. Simiarily the B and C ranges are selected for methane and ethane.

```
B  018..027  [Enter]
C  028..037  [Enter]
```

Now you can select a bar graph with color and view the graph.

```
Type Bar-Graph Options Color Quit View
```

By adding legends and titles a graph like that shown in Figure 18.7 is created.

Graphing Captured Data

You can use the graphing features in 1-2-3 to generate a number of displays of your data. For instrument data the most useful are line plots and bar graphs. Two examples of line plots are shown in Figures 18.8 and 18.9. 1-2-3 graphics provides a standard graphics plotting style so various types of data can be graphed on the same size and type of graph. The graphing program also supports a number of high resolution plotters so high quality graphics and overhead transparencies can be created from your data.

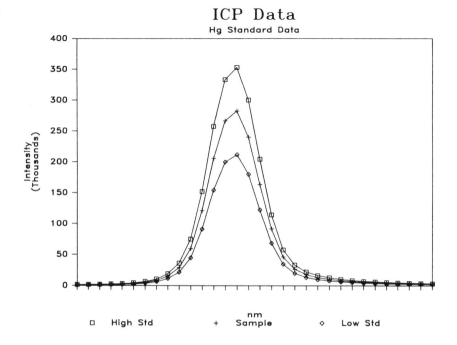

Figure 18.8 Line Plot of ICP Scans of Sample data with high and low calibration data

The data for Figures 18.8 and 18.9 was captured directly from the instruments and stored in a disk file. Then the data was imported into Lotus 1-2-3 with the command:

```
/File Import Numbers
```

This command enters the file contents starting at the current cursor location. For each number in the file 1-2-3 creates a number cell. For each quoted label 1-2-3 creates a left-aligned label. 1-2-3 places successive numbers and labels from the same line in the import file in successive columns of the same row. Thus, a file created with a BASIC program with the contents:

```
"methane",1.27,2534711,400
"ethane",1.88,1183152,350
"propane",2.73,2192368,450
```

is imported into 1-2-3 as:

	A	B	C	D
1	methane	1.27	2534711	400
2	ethane	1.88	1183152	350
3	propane	2.73	2192368	450

The /File Import Number command is very useful for viewing the contents of data files. 1-2-3 extends down to row 8192 so files with as many as 8192 data points can be imported with a single command. For much more on importing data read chapter 3 "Importing Data into 1-2-3."

Numerous
Applications

Once you have used 1-2-3 for a few data management and graphics applications you will think of many more applications for this versatile program. There are a number of major features of the progam which were not even mentioned. For example you can generate Tables of your data using the Data Tables command. Handling data in 1-2-3 is perfect for "first time" interfacing with instruments since the design of the data reduction and display can be performed right in the program. Then if you need special data reduction or display not available in 1-2-3 you can then write a custom program.

19

Frequency Distributions with Lotus 1-2-3

A seldom mentioned Lotus 1-2-3 command has powerful applications in data analysis and laboratory sample management.

Lotus 1-2-3 provides a number of data analysis commands under the Data menu. One of these commands is the Distribute command which is very easy to use and provides extremely valuable results when it is utilized. The command allows you to categorize data values to generate a frequency distribution for the data.

What is a Frequency Distribution?

A frequency distribution is obtained when you sort data and count how many readings fall within selected ranges of value. An example will give you a clear grasp of this method of data analysis.

Let's say we are looking at the Sunday sports section of our newspaper and scan the batting averages of the major league baseball players. These batting averages range from around .360 for the best hitters to around .200. To view how the batting averages are distributed between these extreme values we can select an arbitrary range of say .010 and then count how many players have averages in each range.

Going from low to high the first range would be all below .200. The next range would be from .201 to .210, followed by the range from .211 to .220. These ranges would continue until we have a range from .351 to .360 and finally .361 or greater. Each of these ranges can be thought of as a "bin." By counting the number of players whose batting averages fall in each "bin" we obtain a frequency distribution of the batting averages.

Using Data Distribute in Lotus

We can use Lotus 1-2-3 to generate a frequency distribution on the baseball batting averages. Figure 19.1 shows some of the batting averages along with the player's name and team. The rest of the names and averages continue down the columns and are not shown. In column E, we use the Data Fill command to enter a column of values each .010 apart starting at .200. To do this you would place the cursor at cell E3 and then enter the command:

```
/Data Fill E3..E19 [Enter] .200 [Enter]
     .01 [Enter] .360 [Enter]
```

These values form the "bins" we will use to sort the batting averages into. Now we can perform the sort with the command:

```
/Data Distribute C4..C83 [Enter] E3..E19
[Enter]
```

The first range covers all of the batting averages or data to sort while the second range covers the bin values. The command outputs into the column to the right of the bins the number of values falling in the range "up to and including" that value. In our batting average example, there was one average below .200 and thus there is a one next to the .200. There are two averages above .360 and two between .351 and .360.

We can now plot this frequency distribution with the following Lotus graph commands:

```
/Graph Type Bar X E3..E20 [Enter] A F3..F20
[Enter] View
```

A graph like the one shown in Figure 19.2 will be generated showing the distribution of the batting averages. For a distribution like this to be meaningful, you must wisely select your bin value range and size. Using many small bins or having just a few large bins will produce distributions which will be uninteresting and for analysis purposes unusable.

Laboratory Applications of Frequency Distributions

Frequency distributions can be very valuable for data analysis. A obvious application is to view the distribution of instrument readings. All instrument readings have a certain amount of error. The error should be random and evenly distributed centered around the actual value. You can analyze your own data using the Lotus Data Distribute command. An example of this type of analysis is shown in Figure 19.3. Calibration readings for an instrument on the same calibration standard are entered into the worksheet over a three week time period. The data is held in five columns.

Figure 19.1 Batting
Average Lotus template.
The frequency distribution
of National League Batting
Averages is generated
using this template.

	A	B	C	D	E	F	G
1	National	League	Batting Averages		Through 5 June	1987	
2					Up to:		
3	Player	Team	Average		0.200	1	
4	Krunk	SD	0.377		0.210	1	
5	Gwynn	SD	0.366		0.220	4	
6	Leonard	SF	0.357		0.230	2	
7	Guerrero	LA	0.351		0.240	5	
8	Galarraga	Mon	0.339		0.250	8	
9	Hatcher	Htn	0.332		0.260	8	
10	JClark	StL	0.328		0.270	8	Avg is .278
11	EDavis	Cin	0.327		0.280	6	
12	Maldanald	SF	0.327		0.290	6	
13	Dykstra	NY	0.326		0.300	7	
14	MWilson	NY	0.325		0.310	4	
15	WClark	SF	0.325		0.320	6	
16	DMurphy	Atl	0.323		0.330	7	
17	Griffey	Atl	0.320		0.340	2	
18	Oberkfell	Alt	0.319		0.350	0	
19	Daniels	Cin	0.317		0.360	2	
20	KHernndz	NY	0.317		> .360	2	
21	Hubbard	Atl	0.314				
22	Pndltn	StL	0.314				
23	Herr	StL	0.307				
24	Ashby	Htn	0.306				
25	Durham	Chi	0.304				
26	Candaele	Mon	0.303				
27	Bell	Cin	0.300				
28	Bream	Pit	0.300				
29	Parker	Cin	0.299				
30	Wallach	Mon	0.299				
31	Law	Mon	0.293				
32	Stubbs	LA	0.291				
33	Webster	Mon	0.291				
34	Dawson	Chi	0.290				
35	Scioscia	LA	0.290				
36	Schmidt	Phi	0.285				
37	OSmith	StL	0.284				
38	Sandberg	Chi	0.281				
39	Stllwll	Cin	0.281				
40	GDavis	Htn	0.279				
41	JDavis	Chi	0.277				

42	Bass	Htn	0.275
43	Santiago	SD	0.274
44	Coleman	StL	0.272
45	DJames	Atl	0.271
46	Strawbry	NY	0.270
47	CDavis	SF	0.266
48	Morrison	Pit	0.266
49	DMartinez	Chi	0.265
50	GWilson	Phi	0.264
51	Bonds	Pit	0.263
52	Santana	NY	0.262
53	VanSlyke	Pit	0.262
54	Ray	Pit	0.260
55	Samuel	Phi	0.260
56	Cruz	Htn	0.259
57	McReylds	NY	0.259
58	Sax	LA	0.258
59	BDiaz	Cin	0.254
60	McGee	StL	0.253
61	Hayes	Phi	0.252
62	MThmpsn	Phi	0.250
63	Doran	Htn	0.249
64	Backman	NY	0.248
65	AThomas	Atl	0.246
66	Melvin	SF	0.246
67	Virgil	Atl	0.245
68	Carter	NY	0.243
69	Mitchell	SD	0.242
70	Dunston	Chi	0.239
71	CMartinez	SD	0.238
72	Cora	SD	0.234
73	Ramsey	LA	0.232
74	Oester	Cin	0.231
75	HJohson	NY	0.227
76	GPerry	Atl	0.222
77	Duncan	LA	0.219
78	Parrish	Phi	0.217
79	Moreland	Chi	0.215
80	Belliard	Pit	0.212
81	Templetn	SD	0.209
82	MWilliams	SF	0.191
83			

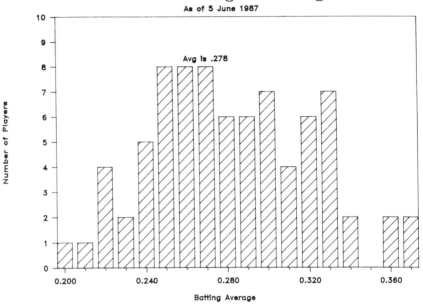

Figure 19.2 Lotus graph showing the frequency distribution of the batting averages. Notice the batting averages are not evenly distributed around the average of .278.

		A	B	C	D	E	F	G	H	I
1	Frequency Distribution Template									
2							Bins			
3	Readings ==>						Up to:	Frequency		
4	Sample	1	2	3	4	5	9.5	1		
5	Calib1	10	10.2	10.1	10.2	9.9	9.6	2		
6	Calib2	9.8	10	10.1	9.7	10.1	9.7	7		
7	Calib3	9.6	9.7	9.6	9.7	9.8	9.8	10		
8	Calib4	9.9	10	9.5	9.7	9.8	9.9	11		
9	Calib5	10.3	10.2	10.1	10.3	10.3	10	15		
10	Calib6	9.9	10	10	9.9	9.8	10.1	14		
11	Calib7	10	10.3	10.1	10.2	9.9	10.2	9		
12	Calib8	9.8	10	10.1	9.7	10.1	10.3	5		
13	Calib9	10.5	10.2	10.1	10.5	10.4	10.4	3		
14	Calib10	9.9	10	10	9.9	9.8	10.5	2		
15	Calib11	10	10.3	9.8	10.2	9.9	10.6	1		
16	Calib12	9.8	10	10.1	9.7	10.1	10.7	0		
17	Calib13	10.4	10.2	10.1	10.6	10.4		0		
18	Calib14	9.9	10	10	9.9	9.8				
19	Calib15	10	10.2	10.1	10.2	9.9				
20	Calib16	9.8	10	10.1	9.7	10.1				

Figure 19.3 Instrument reading template. The frequency distribution for a set of calibration readings over a three week period is analyzed.

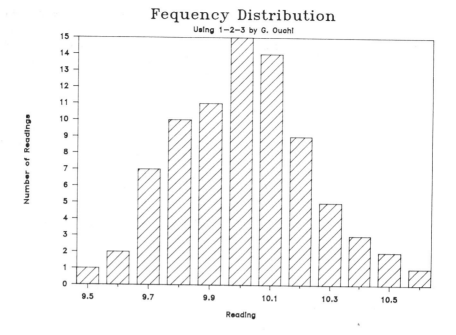

Figure 19.4 Lotus graph of the frequency distribution for the instrument readings. Notice the readings are evenly distributed around the value 10. This type of analysis would be very tedious without a computer and program like Lotus 1-2-3.

We first use the Data Fill command to create our bin values. Place the cursor at H4 and enter the command:

```
/Data Fill H4..H16 [Enter] 9.5 [Enter]
.1 [Enter] 10.7 [Enter]
```

Then generate the distribution with the command:

```
/Data Distribute C5..G20 [Enter] H4..H16
[Enter]
```

We can then view the distribution as shown in Figure 19.4 with the commands:

```
/Graph Type Bar X H4..H15 [Enter] A I4..I15
[Enter] View
```

With a graph of this type you can easily view the distribution of your data and see whether it is distributed evenly or it is being skewed by an unknown factor. This type of analysis is essential for instruments like microtiter plate readers which perform hundreds of analyses. These instruments should be checked for accuracy and even distribution of readings on a regular basis.

Frequency of Analysis

A common frequency distribution needed by laboratory managers is the frequency of analysis. This type of sample trend analysis will answer questions like: How many samples were analyzed last month ? How many GC samples did we run last week? How many samples did we run for Joe Jones in June?

To track this data, we need to create a small database like the one shown in Figure 19.5. This database has a sample number in column A, the submitters name in column B, the date the analysis was performed and the analysis type in columns C and D. Dates are entered into the worksheet using the Lotus DATE function. This function generates a number which is the number of days since January 1, 1900. This value can then be displayed in one of many date formats by using the /Range Format Date command. The DATE function has the following structure:

```
@DATE(87,5,1)

(Year, Month, Day)
```

	A	B	C	D	E	F	G
1	Laboratory Sample Log						
2							
3						Bin	Frequency
4	Sample	Submitter	Date	Analysis		Week Ending	
5	20342	M. Johnson	04-May-87	GCHROM		01-May-87	0
6	20343	Bird	07-May-87	LCHROM		08-May-87	3
7	20344	McHale	07-May-87	LCHROM		15-May-87	4
8	20345	Ainge	12-May-87	UV		22-May-87	6
9	20346	Parish	14-May-87	GCHROM		29-May-87	3
10	20347	D. Johnson	16-May-87	LCHROM		05-Jun-87	0
11	20348	Cooper	16-May-87	LCHROM			0
12	20349	Worthy	18-May-87	UV			
13	20350	Abdul-Jabbar	19-May-87	IR			
14	20351	Green	20-May-87	LCHROM			
15	20352	Scott	21-May-87	LCHROM			
16	20353	Thompson	22-May-87	UV			
17	20354	Rambis	22-May-87	IR			
18	20355	Kite	26-May-87	UV			
19	20356	Daye	25-May-87	GCHROM			
20	20357	Sichting	25-May-87	NMR			

Figure 19.5 Laboratory Sample Log used to track the frequency of samples analyzed in this lab. Each sample is logged in when the analysis is completed. The frequency distribution shows how many samples were analyzed each week.

For these parameters, the function will generate the number 31898. This value can then be displayed on the worksheet as:

```
1-May-87
```

by formatting the cell with the command:

```
/Range Format Date 1 cell-range
```

Using the DATE function makes handling dates very easy. It is easy to find the number of days between two dates by simply taking the difference between the values generated by the DATE function. You can also use these values when you perform data management queries within Lotus. They can be used with any commands or functions which use numbers.

Seeing the Analysis Distribution

To see the number of samples analyzed during each week we can again use the Data Distribute command. This time our bin values are created using the DATE function. We want to have the number of samples analyzed each week. The bin values are generated as follows. Refering to Figure 19.5, in cell F5 we enter the function @DATE (87,5,1) to get the value for 1 May 1987. In cell F6 we enter the formula +F5+7. This will create the value for the day seven days after May 1, 1987. We can now copy this formula to the other cells with the command:

```
/Copy F6 [Enter] F7..F10 [Enter]
```

The dates are then displayed with the command:

```
/Range Format Date 1 F5..F10 [Enter]
```

Now we can perform the distribution with the command:

```
/Data Distribute C5..C20 [Enter] F5..F10
[Enter]
```

The frequency distribution then shows how many samples were analyzed during each week.

Expanding the Worksheet

This worksheet could be sorted and the data distribution range could be selected to include only specific analysis types or submitters. Then you would see the frequency of analysis for these special groups. You could expand your database to include the date the sample was submitted and then track the number of days it takes to analyze samples.

The Lotus Data Distribute command has many more applications waiting for you to discover. It can be a very valuable analysis tool to help you analyze your data quality and manage your lab.

Appendix A

Table of ASCII character codes used on IBM PCs and compatibles. To get the binary codes for a character substitute the following binary codes for each of the letters or numbers in the Hex column:

0 = 0000	4 = 0100	8 = 1000	C = 1100
1 = 0001	5 = 0101	9 = 1001	D = 1101
2 = 0010	6 = 0110	A = 1010	E = 1110
3 = 0011	7 = 0111	B = 1011	F = 1111

For example, the capital letter R: hex = 52 or 0101 0010

Dec	Hex	Chr		Dec	Hex	Chr		Dec	Hex	Chr		Dec	Hex	Chr
Ø	ØØ			32	20			64	40	@		96	60	`
1	Ø1			33	21	!		65	41	A		97	61	a
2	Ø2			34	22	"		66	42	B		98	62	b
3	Ø3	♥		35	23	#		67	43	C		99	63	c
4	Ø4	♦		36	24	$		68	44	D		100	64	d
5	Ø5	♣		37	25	%		69	45	E		101	65	e
6	Ø6	♠		38	26	&		70	46	F		102	66	f
7	Ø7	•		39	27	'		71	47	G		103	67	g
8	Ø8			40	28	(		72	48	H		104	68	h
9	Ø9	○		41	29	)		73	49	I		105	69	i
10	ØA			42	2A	*		74	4A	J		106	6A	j
11	ØB	♂		43	2B	+		75	4B	K		107	6B	k
12	ØC	♀		44	2C	,		76	4C	L		108	6C	l
13	ØD	♪		45	2D	–		77	4D	M		109	6D	m
14	ØE	♫		46	2E	.		78	4E	N		110	6E	n
15	ØF	☼		47	2F	/		79	4F	O		111	6F	o
16	10	►		48	30	Ø		80	50	P		112	70	p
17	11	◄		49	31	1		81	51	Q		113	71	q
18	12	↕		50	32	2		82	52	R		114	72	r
19	13	‼		51	33	3		83	53	S		115	73	s
20	14	¶		52	34	4		84	54	T		116	74	t
21	15	§		53	35	5		85	55	U		117	75	u
22	16	▬		54	36	6		86	56	V		118	76	v
23	17	↨		55	37	7		87	57	W		119	77	w
24	18	↑		56	38	8		88	58	X		120	78	x
25	19	↓		57	39	9		89	59	Y		121	79	y
26	1A	→		58	3A	:		90	5A	Z		122	7A	z
27	1B	←		59	3B	;		91	5B	[		123	7B	{
28	1C	∟		60	3C	<		92	5C	\		124	7C		
29	1D	↔		61	3D	=		93	5D	]		125	7D	}
30	1E	▲		62	3E	>		94	5E	^		126	7E	~
31	1F	▼		63	3F	?		95	5F	—		127	7F	⌂

(continued)

Dec	Hex	Chr	Dec	Hex	Chr	Dec	Hex	Chr	Dec	Hex	Chr
128	80	Ç	160	A0	á	192	C0	└	224	E0	α
129	81	ü	161	A1	í	193	C1	┴	225	E1	β
130	82	é	162	A2	ó	194	C2	┬	226	E2	Γ
131	83	â	163	A3	ú	195	C3	├	227	E3	π
132	84	ä	164	A4	ñ	196	C4	─	228	E4	Σ
133	85	à	165	A5	Ñ	197	C5	┼	229	E5	σ
134	86	å	166	A6	ª	198	C6	╞	230	E6	µ
135	87	ç	167	A7	º	199	C7	╟	231	E7	τ
136	88	ê	168	A8	¿	200	C8	╚	232	E8	Φ
137	89	ë	169	A9	⌐	201	C9	╔	233	E9	Θ
138	8A	è	170	AA	¬	202	CA	╩	234	EA	Ω
139	8B	ï	171	AB	½	203	CB	╦	235	EB	δ
140	8C	î	172	AC	¼	204	CC	╠	236	EC	∞
141	8D	ì	173	AD	¡	205	CD	═	237	ED	φ
142	8E	Ä	174	AE	«	206	CE	╬	238	EE	ε
143	8F	Å	175	AF	»	207	CF	╧	239	EF	∩
144	90	É	176	B0	░	208	D0	╨	240	F0	≡
145	91	æ	177	B1	▒	209	D1	╤	241	F1	±
146	92	Æ	178	B2	▓	210	D2	╥	242	F2	≥
147	93	ô	179	B3	│	211	D3	╙	243	F3	≤
148	94	ö	180	B4	┤	212	D4	╘	244	F4	⌠
149	95	ò	181	B5	╡	213	D5	╒	245	F5	⌡
150	96	û	182	B6	╢	214	D6	╓	246	F6	÷
151	97	ù	183	B7	╖	215	D7	╫	247	F7	≈
152	98	ÿ	184	B8	╕	216	D8	╪	248	F8	°
153	99	Ö	185	B9	╣	217	D9	┘	249	F9	∙
154	9A	Ü	186	BA	║	218	DA	┌	250	FA	·
155	9B	¢	187	BB	╗	219	DB	█	251	FB	√
156	9C	£	188	BC	╝	220	DC	▄	252	FC	ⁿ
157	9D	¥	189	BD	╜	221	DD	▌	253	FD	²
158	9E	₧	190	BE	╛	222	DE	▐	254	FE	■
159	9F	ƒ	191	BF	┐	223	DF	▀	255	FF	

Index